AMS SHORT COURSE LECTURE NOTES

Introductory Survey Lectures

published as a subseries of

Proceedings of Symposia in Applied Mathematics

Proceedings of Symposia in
APPLIED MATHEMATICS

Volume 69

Evolutionary Game Dynamics

American Mathematical Society
Short Course
January 4–5, 2011
New Orleans, Louisiana

Karl Sigmund
Editor

American Mathematical Society
Providence, Rhode Island

2010 *Mathematics Subject Classification.* Primary 91A22.

Library of Congress Cataloging-in-Publication Data

American Mathematical Society. Short Course (2011 : New Orleans, LA.)
Evolutionary game dynamics : American Mathematical Society Short Course, January 4–5, 2011, New Orleans, LA / Karl Sigmund, editor.
p. cm. — (Proceedings of symposia in applied mathematics ; v. 69)
Includes bibliographical references and index.
ISBN 978-0-8218-5326-9 (alk. paper)
1. Game theory—Congresses. I. Sigmund, Karl, 1945– II. Title.

QA269.A465 2011
519.3—dc23

2011028869

♾ The paper used in this book is acid-free and falls within the guidelines
established to ensure permanence and durability.
Visit the AMS home page at http://www.ams.org/

10 9 8 7 6 5 4 3 2 1 16 15 14 13 12 11

Contents

Preface

Evolutionary game theory studies basic types of social interactions in populations of players. It is the ideal mathematical tool for methodological individualism, i.e., the reduction of social phenomena to the level of individual actions. Evolutionary game dynamics combines the strategic viewpoint of classical game theory (independent, rational players trying to outguess each other) with population dynamics (successful strategies increase their frequencies).

A substantial part of the appeal of evolutionary game theory comes from its highly diverse applications, such as social dilemmas, the evolution of language, or mating behavior in animals. Moreover, its methods are becoming increasingly popular in computer science, engineering, and control theory. They help to design and control multi-agent systems, often with large number of agents (for instance, when routing drivers over highway networks, or data packets over the Internet). While traditionally these fields have used a top down approach, by directly controlling the behavior of each agent in the system, attention has recently turned to an indirect approach: allowing the agents to function independently, while providing incentives that lead them to behave in the desired way. Instead of the traditional assumption of equilibrium behavior, researchers opt increasingly for the evolutionary paradigm, and consider the dynamics of behavior in populations of agents employing simple, myopic decision rules. The methods of evolutionary game theory are used in disciplines as diverse as microbiology, genetics, animal behavior, evolutionary psychology, route planning, e-auctions, common resources management or micro-economics.

The present volume is based on a mini-course held at the AMS meeting in New Orleans in January 2011. The lectures deal mostly with the mathematical aspects of evolutionary game theory, i.e., with the deterministic and stochastic dynamics describing the evolution of frequencies of behavioral types.

An introductory part of the course is devoted to a brief sketch of the origins of the field, and in particular to the examples that motivated evolutionary biologists to introduce a population dynamical viewpoint into game theory. This leads to some of the main concepts: evolutionary stability, replicator dynamics, invasion fitness, etc. Much of it can be explained by means of simple examples such as the Rock-Paper-Scissors game. It came as a surprise when childish games of that sort, intended only for the clarification of concepts, were found to actually lurk behind important classes of real-life social and biological interactions. The transmission of successful strategies by genetic and cultural means results in a rich variety of stochastic processes and, in the limit of very large populations, deterministic adjustment dynamics including differential inclusions and reaction-diffusion equations.

Some economists view these types of dynamics merely as tools for so-called equilibrium refinement and equilibrium selection concepts. (Indeed, most games have so many equilibria that it is hard to select the 'right one'). However, evolutionary games have also permitted us to move away from the equilibrium-centered viewpoint. Today, we understand that it is often premature to assume that behavior converges to an equilibrium. In particular, an evolutionarily stable strategy need not be reachable. A homogeneous population using that strategy cannot be invaded by a minority of dissidents, but a homogeneous population with a slightly different strategy can evolve away from it. Limit phenomena such as periodic or heteroclinic cycles, or chaotic attractors, may be considered, perhaps not as 'solutions of the game', but as predictions of play. On the other hand, large classes of games leading to global convergence are presently much better understood.

This book offers a succinct state-of-the-art introduction to the increasingly sophisticated mathematical techniques behind evolutionary game theory.

Proceedings of Symposia in Applied Mathematics
Volume **69**, 2011

Introduction to Evolutionary Game Theory

Karl Sigmund

ABSTRACT. This chapter begins with some basic terminology, introducing elementary game theoretic notions such as payoff, strategy, best reply, Nash equilibrium pairs etc. Players who use strategies which are in Nash equilibrium have no incentive to deviate unilaterally. Next, a population viewpoint is introduced. Players meet randomly, interact according to their strategies, and obtain a payoff. This payoff determines how the frequencies in the strategies will evolve. Successful strategies spread, either (in the biological context) through inheritance or (in the cultural context) through social learning. The simplest description of such an evolution is based on the replicator equation. The basic properties of replicator dynamics are analyzed, and some low-dimensional examples such as the Rock-Scissors-Paper game are discussed. The relation between Nash equilibria and rest points of the replicator equation are investigated, which leads to a short proof of the existence of Nash equilibria. We then study mixed strategies and evolutionarily stable strategies. This introductory chapter continues with a brief discussion of other game dynamics, such as the best reply dynamics, and ends with the simplest extension of replicator dynamics to asymmetric games.

1. Predictions and Decisions

Predictions can be difficult to make, especially, as Niels Bohr quipped, if they concern the future. Reliable forecasts about the weather or about some social development may seem to offer comparable challenges, at first sight. But there is a fundamental difference: a weather forecast does not influence the weather, whereas a forecast on economy can influence the economic outcome. Humans will react if they learn about the predictions, and they can anticipate that others will react, too.

When the economist Oskar Morgenstern, in the early 'thirties, became aware of the problem, he felt that he had uncovered an 'impossibility theorem' of a similarly fundamental nature as the incompleteness theorem of his friend, the logician Kurt Gödel. Morgenstern was all the more concerned about it as he was director of the Vienna-based Institut für Konjunkturforschung, the Institute for Business Cycles Research, whose main task was actually to deliver predictions on the Austrian economy. Oscar Morgenstern expained his predicament in many lectures and publications, using as his favorite example the pursuit of Sherlock Holmes by the

2000 *Mathematics Subject Classification.* Primary 91A22.
The author wants to thank Ross Cressman for his helpful comments.

infamous Professor Moriarty [**24**]. These two equally formidable adversaries would never arrive at a conclusive solution in mutually outguessing each other.

We can describe the fundamental nature of the problem by using some of the mathematical notation which later was introduced through game theory. Let us suppose that player I has to choose between n options, or *strategies*, which we denote by $\mathbf{e}_1,..., \mathbf{e}_n$, and player II between m strategies $\mathbf{f}_1,..., \mathbf{f}_m$. If I chooses $\mathbf{e}_i$ and II chooses $\mathbf{f}_j$, then player I obtains a *payoff* a_{ij} and player II obtains b_{ij}. The game, then, is described by two $n \times m$ payoff matrices A and B: alternatively, we can describe it by one matrix whose element, in the i-th row and j-th column, is the pair (a_{ij}, b_{ij}) of payoff values. The payoff is measured on a utility scale consistent with the players' preference ranking.

The two players could engage in the game 'Odd or Even?' and decide that the loser pays one dollar to the winner. At a given signal, each player holds up one or two fingers. If the resulting sum is odd, player I wins. If the sum is even, player II wins. Each player then has to opt for *even* and *odd*, which correspond to $\mathbf{e}_1$ and $\mathbf{e}_2$ for player I and $\mathbf{f}_1$ and $\mathbf{f}_2$ for player II, and the payoff matrix is

$$\begin{pmatrix} (-1,1) & (1,-1) \\ (1,-1) & (-1,1) \end{pmatrix} \tag{1.1}$$

If the two players graduate to the slightly more sophisticated Rock-Scissors-Paper game, they would each have to opt between three strategies, numbered in that order, and the payoff matrix would be

$$\begin{pmatrix} (0,0) & (1,-1) & (-1,1) \\ (-1,1) & (0,0) & (1,-1) \\ (1,-1) & (-1,1) & (0,0) \end{pmatrix}. \tag{1.2}$$

If both Rock-Scissors-Paper players opt for the same move, the game is a tie and both obtain payoff zero. If the outcome is $(0,0)$ or $(-1,1)$, then player I (who chooses the row of the payoff matrix) would have done better to choose another strategy; if the outcome is $(1,-1)$ or $(0,0)$, then it is player II, the column player, who would have done better to switch. If a prediction is made public, then at least one of the players would have an incentive to deviate. The other player would anticipate this, and deviate accordingly, and both would be launched into a vicious circle of mutual outguessing.

A few years, however, after Morgenstern had started to broadcast his impossibility result, the topologist Cech pointed out to him that John von Neumann had found, in an earlier paper on parlor games, a way to avoid Morgenstern's dead end [**42**]. It consists in randomizing, i.e. letting chance decide. Clearly, if players opt with equal probability for each of their alternatives, none has an incentive to deviate. Admittedly, this would lead to the expected payoff 0, somewhat of an anti-climax. But John von Neumann's minimax theorem holds for a much larger class of games. Most importantly, it led, in the 'forties, to a collaboration of John von Neumann with Oscar Morgenstern which gave birth to game theory [**43**]. A few years later, John Nash introduced an equilibrium notion valid in an even more general context, which became the cornerstone of game theory [**33**].

2. Mixed strategies and best replies

Suppose that player I opts to play strategy $\mathbf{e}_i$ with probability x_i. This *mixed* strategy is thus given by a stochastic vector $\mathbf{x} = (x_1, ..., x_n)$ (with $x_i \geq 0$ and

$x_1 + ... + x_n = 1$). We denote the set of all such mixed strategies by Δ_n: this is a simplex in R^n, spanned by the unit vectors $\mathbf{e}_i$ of the standard base, which are said to be the *pure* strategies, and correspond to the original set of alternatives.

Similarly, a mixed strategy for player II is an element $\mathbf{y}$ of the unit simplex Δ_m spanned by the unit vectors $\mathbf{f}_j$. If player I uses the pure strategy $\mathbf{e}_i$ and player II uses strategy $\mathbf{y}$, then the payoff for player I (or more precisely, its expected value) is

$$(A\mathbf{y})_i = \sum_{j=1}^{m} a_{ij} y_j. \tag{2.1}$$

If player I uses the mixed strategy $\mathbf{x}$, and II uses $\mathbf{y}$, the payoff for player I is

$$\mathbf{x} \cdot A\mathbf{y} = \sum_i x_i (A\mathbf{y})_i = \sum_{i,j} a_{ij} x_i y_j, \tag{2.2}$$

and the payoff for player II, similarly, is

$$\mathbf{x} \cdot B\mathbf{y} = \sum_{i,j} b_{ij} x_i y_j. \tag{2.3}$$

If player I knows the strategy $\mathbf{y}$ of the co-player, then player I should use a strategy which is a *best reply* to $\mathbf{y}$. The set of best replies is the set

$$BR(\mathbf{y}) = \arg\max_{\mathbf{x}} \mathbf{x} \cdot A\mathbf{y}, \tag{2.4}$$

i.e. the set of all $\mathbf{x} \in \Delta_n$ such that $\mathbf{z} \cdot A\mathbf{y} \leq \mathbf{x} \cdot A\mathbf{y}$ holds for all $\mathbf{z} \in \Delta_n$. Player I has no incentive to deviate from $\mathbf{x}$ and chose another strategy $\mathbf{z}$ instead.

Since the function $\mathbf{z} \mapsto \mathbf{z} \cdot A\mathbf{y}$ is continuous and Δ_n is compact, the set of best replies is always non-empty. It is a convex set. Moreover, if $\mathbf{x}$ belongs to $BR(\mathbf{y})$, so do all pure strategies in the *support* of $\mathbf{x}$, i.e. all $\mathbf{e}_i$ for which $x_i > 0$. Indeed, for all i,

$$(A\mathbf{y})_i = \mathbf{e}_i \cdot A\mathbf{y} \leq \mathbf{x} \cdot A\mathbf{y}. \tag{2.5}$$

If the inequality sign were strict for some i with $x_i > 0$, then $x_i(A\mathbf{y})_i < x_i(\mathbf{x} \cdot A\mathbf{y})$; summing over all $i = 1, ..., n$ then leads to a contradiction. It follows that the set $BR(\mathbf{y})$ is a face of the simplex Δ_n. It is spanned by the pure strategies which are best replies to $\mathbf{y}$.

If player I has found a best reply to the strategy $\mathbf{y}$ of player II, then player I has no incentive not to use it, as long as player II sticks to $\mathbf{y}$. But will player II stick to $\mathbf{y}$? Only if player II has no incentive either to use another strategy, i.e. has also hit upon a best reply. Two strategies $\mathbf{x}$ and $\mathbf{y}$ are said to form a *Nash equilibrium* pair if each is a best reply to the other, i.e., if $\mathbf{x} \in BR(\mathbf{y})$ and $\mathbf{y} \in BR(\mathbf{x})$, or alternatively if

$$\mathbf{z} \cdot A\mathbf{y} \leq \mathbf{x} \cdot A\mathbf{y} \tag{2.6}$$

holds for all $\mathbf{z} \in \Delta_n$, and

$$\mathbf{x} \cdot B\mathbf{w} \leq \mathbf{x} \cdot B\mathbf{y} \tag{2.7}$$

holds for all $\mathbf{w} \in \Delta_m$. A Nash equilibrium pair $(\mathbf{x}, \mathbf{y})$ satisfies a minimal consistency requirement: no player has an incentive to deviate (as long as the other player does not deviate either).

A basic result states that there always exist Nash equilibrium pairs, for any game (A, B). The result holds for vastly wider classes of games than considered so far; it holds for any number of players, any convex compact sets of strategies, any continuous payoff functions, and even beyond (see, e.g., [**30**]). But it would not hold if we had not allowed for mixed strategies, as is shown by the Rock-Scissors-Paper game. In that case, the mixed strategy which consists in choosing, with equal probability 1/3, among the three alternative moves, clearly leads to an equilibrium pair. No player has a reason to deviate. On the other hand, if player I uses any other strategy (x_1, x_2, x_3) against the $(1/3, 1/3, 1/3)$ of player II, player I would still have an expected payoff of 0. However, the *other* player II would then have an incentive to deviate, presenting I with an incentive to deviate in turn, and so on.

In this example, $(\mathbf{x}, \mathbf{y})$ with $\mathbf{x} = \mathbf{y} = (1/3, 1/3, 1/3)$ is the unique Nash equilibrium pair. We have seen that as long as player II chooses the equilibrium strategy $\mathbf{y}$, player I has no reason to deviate from the equilibrium strategy $\mathbf{x}$, but that on the other hand, player I has no reason not to deviate, either. This would be different if $(\mathbf{x}, \mathbf{y})$ were a *strict* Nash equilibrium pair, i.e. if

$$\mathbf{z} \cdot A\mathbf{y} < \mathbf{x} \cdot A\mathbf{y} \tag{2.8}$$

holds for all $\mathbf{z} \neq \mathbf{x}$, and

$$\mathbf{x} \cdot B\mathbf{w} < \mathbf{x} \cdot B\mathbf{y} \tag{2.9}$$

holds for all $\mathbf{w} \neq \mathbf{y}$. In this case, i.e. when both best-reply sets are singletons, and hence correspond to pure strategies, each player will be penalized for unilaterally deviating from the equilibrium.

Whereas every game admits a Nash equilibrium pair, some need not admit a strict Nash equilibrium pair, as our previous examples show.

3. Excurse to zero sum

Historically, game theory focused first on zero-sum games, for which $a_{ij} = -b_{ij}$ for all i, j, i.e., $A = -B$ (the gain of player I is the loss of player II). This condition clearly holds for a large set of parlor games. But it certainly restricts the range of applications. For most types of social and economic interactions, the assumption that the interests of the two players are always diametrically opposite does not hold. Even in military confrontations, there often exist outcomes both parties want to avoid. Most interactions are of mixed motive type, and contain elements of cooperation as well as competition. Interestingly, John von Neumann did not greatly appreciate the solution concept proposed by the undergraduate student John Nash. A short interview ended when John von Neumann remarked, somewhat dismissively, 'Oh, just a fixed point theorem' [**32**]. We shall see that the existence proof for Nash equilibrium pairs does indeed reduce to a fixed point theorem, and a rather simple one at that. Nevertheless, it yields a very powerful result, as can be seen by applying it to the special case of zero sum games, where it leads to a three-liner proof of the celebrated maximin theorem, a proof which is considerably simpler than John von Neumann's original, brute-force demonstration.

It is easy to see that $(\bar{\mathbf{x}}, \bar{\mathbf{y}})$ is a Nash equilibrium pair of a zero-sum game iff

$$\mathbf{x} \cdot A\bar{\mathbf{y}} \leq \bar{\mathbf{x}} \cdot A\bar{\mathbf{y}} \leq \bar{\mathbf{x}} \cdot A\mathbf{y} \tag{3.1}$$

for all $\mathbf{x} \in \Delta_n, \mathbf{y} \in \Delta_m$. Suppose that player II correctly guesses that player I plays $\mathbf{x}$. Then player II will use a strategy which is a best reply, i.e., minimizes player I's payoff, which will reduce to $g(\mathbf{x}) := \min_{\mathbf{y}} \mathbf{x} \cdot A\mathbf{y}$. A player I who expects to be anticipated, then, ought to maximize $g(\mathbf{x})$. Any strategy $\hat{\mathbf{x}}$ yielding this maximum is said to be a maximin strategy for player I. Such a maximin strategy is defined by $\hat{\mathbf{x}} := \arg\max_{\mathbf{x}} g(\mathbf{x})$, and it guarantees player I a security level

$$w_u := \max_{\mathbf{x}} \min_{\mathbf{y}} \mathbf{x} \cdot A\mathbf{y}. \tag{3.2}$$

Similarly, we can expect player II to maximize the own security level, i.e., since $A = -B$, to play a minimax strategy $\hat{\mathbf{y}}$ such that player I has a payoff bounded from above by

$$w_o := \min_{\mathbf{y}} \max_{\mathbf{x}} \mathbf{x} \cdot A\mathbf{y}. \tag{3.3}$$

The pair $(\hat{\mathbf{x}}, \hat{\mathbf{y}})$ is said to be a *maximin pair.* It satisfies

$$\min_{\mathbf{y}} \hat{\mathbf{x}} \cdot A\mathbf{y} = w_u, \qquad \max_{\mathbf{x}} \mathbf{x} \cdot A\hat{\mathbf{y}} = w_o, \tag{3.4}$$

and leads to a payoff which clearly satisfies

$$w_u \le \hat{\mathbf{x}} \cdot A\hat{\mathbf{y}} \le w_o. \tag{3.5}$$

If $(\bar{\mathbf{x}}, \bar{\mathbf{y}})$ is a Nash equilibrium pair for a zero sum game, then it is a maximin pair. Indeed, by (3.1),

$$\max_{\mathbf{x}} \mathbf{x} \cdot A\bar{\mathbf{y}} \le \bar{\mathbf{x}} \cdot A\bar{\mathbf{y}} \le \min_{\mathbf{y}} \bar{\mathbf{x}} \cdot A\mathbf{y}. \tag{3.6}$$

Now by (3.3), w_o is less than the left hand side of the previous inequality and by (3.2) w_u larger than the right hand side. Since $w_u \le w_o$ by (3.5), we must actually have equality everywhere. But $w_u = \min_{\mathbf{y}} \bar{\mathbf{x}} \cdot A\mathbf{y}$ means that $\bar{\mathbf{x}}$ is a maximin solution, and $\max_{\mathbf{x}} \mathbf{x} \cdot A\bar{\mathbf{y}} = w_o$ that $\bar{\mathbf{y}}$ is a minimax solution.

For zero-sum games, the existence of a Nash equilibrium pair thus implies the existence of a maximin pair. The previous argument implies $w_u = w_o$, i.e.,

$$\min_{\mathbf{y}} \max_{\mathbf{x}} \mathbf{x} \cdot A\mathbf{y} = \max_{\mathbf{x}} \min_{\mathbf{y}} \mathbf{x} \cdot A\mathbf{y}. \tag{3.7}$$

Conversely, it is easy to see that if $(\hat{\mathbf{x}}, \hat{\mathbf{y}})$ is a maximin pair of a zero sum game, then it is a Nash equilibrium pair.

4. Concerns about the Nash solution

Let us note that if $(\hat{\mathbf{x}}, \hat{\mathbf{y}})$ and $(\bar{\mathbf{x}}, \bar{\mathbf{y}})$ are two Nash equilibrium pairs for a zero sum game, then so are $(\hat{\mathbf{x}}, \bar{\mathbf{y}})$ and $(\bar{\mathbf{x}}, \hat{\mathbf{y}})$. Indeed,

$$\hat{\mathbf{x}} \cdot A\hat{\mathbf{y}} \le \hat{\mathbf{x}} \cdot A\bar{\mathbf{y}} \le \bar{\mathbf{x}} \cdot A\bar{\mathbf{y}} \le \bar{\mathbf{x}} \cdot A\hat{\mathbf{y}} \le \hat{\mathbf{x}} \cdot A\hat{\mathbf{y}}, \tag{4.1}$$

hence equality holds everywhere, and therefore for all $\mathbf{x}$ and $\mathbf{y}$:

$$\mathbf{x} \cdot A\bar{\mathbf{y}} \le \hat{\mathbf{x}} \cdot A\bar{\mathbf{y}} \le \hat{\mathbf{x}} \cdot A\mathbf{y} \tag{4.2}$$

so that $(\hat{\mathbf{x}}, \bar{\mathbf{y}})$ is a Nash equilibrium pair.

The same need not hold for general (non zero-sum) games. Consider for instance

$$\begin{pmatrix} (1,1) & (-1,-1) \\ (-1,-1) & (1,1) \end{pmatrix} \tag{4.3}$$

It is easy to see that $(\mathbf{e}_1, \mathbf{f}_1)$ and $(\mathbf{e}_2, \mathbf{f}_2)$ are two Nash equilibrium pairs. But $(\mathbf{e}_1, \mathbf{f}_2)$ or $(\mathbf{e}_2, \mathbf{f}_1)$ are not. How should the two players coordinate their choice? The problem becomes even more acute for a coordination game given by

$$\begin{pmatrix} (2,2) & (-100,0) \\ (0,-100) & (1,1) \end{pmatrix}. \tag{4.4}$$

Again, $(\mathbf{e}_1, \mathbf{f}_1)$ and $(\mathbf{e}_2, \mathbf{f}_2)$ are two Nash equilibrium pairs. The former has the advantage of yielding a higher payoff to both players: it is said to be *Pareto-optimal.* But the second is less risky, and therefore said to be *risk-dominant.* Indeed, it can be very costly to go for the Pareto-optimum if the other player fails to do so. It may actually be best to decide against using the Pareto-optimum right away. In any case, if the game is not zero-sum, Nash equilibrium pairs may not offer much help for decision makers.

Moreover, even if there exists a unique Nash equilibrium pair, it can lead to frustration, as in the following example:

$$\begin{pmatrix} (10,10) & (-5,15) \\ (15,-5) & (0,0) \end{pmatrix}. \tag{4.5}$$

In this case, $\mathbf{e_2}$ is the best reply to every (pure or mixed) strategy of player II, and similarly $\mathbf{f}_2$ is always the best reply for player II. Hence $(\mathbf{e}_2, \mathbf{f}_2)$ is the unique Nash equilibrium pair, and it is strict. This game is an example of a *Prisoner's Dilemma game.* The payoff matrix may occur, for instance, if two players are asked to choose, independently and anonymously, whether or not to provide a gift of 15 dollars to the co-player, at a cost of 5 dollars to themselves. It the two players cooperate by both opting for their first strategy, they will end up with 10 dollars each. But each has an incentive to deviate. It is only when both opt for their second solution and defect, that they cannot do better by choosing to deviate. But then, they end up with zero payoff. Let us remark that this dilemma cannot be solved by appealing to non-monetary motivations. It holds whenever the payoff values reflect each players' preference ordering, which may well include a concern for the other.

5. Population Games

So far, we have considered games between two specific players trying to guess each other's strategy and find a best reply. This belongs to the realm of classical game theory, and leads to interesting mathematical and economic developments. Starting with the 'sixties and 'seventies, both theory and applications were considerably stimulated by problems in evolutionary biology, such as sex-ratio theory or the investigation of fighting behavior [**12, 27**]. It required a radical shift in perspective and the introduction of thinking in terms of populations [**29**]. It provided a welcome tool for the analysis of frequency dependent selection and, later, of learning processes.

Let us therefore consider a *population* of players, each with a given strategy. From time to time, two players meet randomly and play the game, using their strategies. We shall consider these strategies as behavioral programs. Such programs can be learned, or inherited, or imprinted in any other way. In a biological setting, strategies correspond to different types of individuals (or behavioral phenotypes). The outcome of each encounter yields payoff values which are no longer measured on utility scales reflecting the individual preferences of the players, but in the one currency that counts in Darwinian evolution, namely fitness, i.e., average reproductive success. If we assume that strategies can be passed on to the offspring, whether through inheritance or through learning, then we can assume that more successful strategies spread.

In order to analyze this set-up, it is convenient to assume, in a first approach, that all individuals in the population are indistinguishable, except in their way of interacting, i.e. that the players differ only by their strategy. This applies well to games where both players are on an equal footing. Admittedly, there are many examples of social interactions which display an inherent asymmetry between the two players: for instance, between buyers and sellers, or between parents and offspring. We will turn to such interactions later.

Thus we start by considering only *symmetric* games. In the case of two-player games, this means that the game remains unchanged if I and II are permuted. In particular, the two players have the same set of strategies. Hence we assume that $n = m$ and $\mathbf{f}_j = \mathbf{e}_j$ for all j; and if a player plays strategy $\mathbf{e}_i$ against someone using strategy $\mathbf{e}_j$ (which is the former $\mathbf{f}_j$), then that player receives the same payoff, whether labeled I or II. Hence $a_{ij} = b_{ji}$, the payoff for a $\mathbf{e}_i$-player against a $\mathbf{e}_j$-players does not depend on who is labelled I and who is II, or in other words $B = A^T$. Thus a symmetric game is specified by the pair (A, A^T), and therefore is defined by a single, square payoff matrix A. All examples encountered so far are symmetric, with the exception of 'Even or Odd'. A zero-sum game which is symmetric must satisfy $A^T = -A$ and hence corresponds to a skew-symmetric payoff matrix.

It is easy to see that the symmetric game given by

$$\begin{pmatrix} -1 & 1 \\ 1 & -1 \end{pmatrix}, \tag{5.1}$$

where success depends on doing the opposite of the co-player, admits $(\mathbf{e}_1, \mathbf{e}_2)$ and $(\mathbf{e}_2, \mathbf{e}_1)$ as asymmetric Nash equilibrium pairs. These are plainly irrelevant as solutions of the game, since it is impossible to distinguish players I and II. Of interest are only symmetric Nash equilibrium pairs, i.e. pairs of strategies $(\mathbf{x}, \mathbf{y})$ with $\mathbf{x} = \mathbf{y}$. A *symmetric Nash equilibrium*, thus, is specified by *one* strategy $\mathbf{x}$ having the property that it is a best reply to itself (i.e. $\mathbf{x} \in BR(\mathbf{x})$). In other words, we must have

$$\mathbf{z} \cdot A\mathbf{x} \leq \mathbf{x} \cdot A\mathbf{x} \tag{5.2}$$

for all $\mathbf{z} \in \Delta_n$. A symmetric *strict* Nash equilibrium is accordingly given by the condition

$$\mathbf{z} \cdot A\mathbf{x} < \mathbf{x} \cdot A\mathbf{x} \tag{5.3}$$

for all $\mathbf{z} \neq \mathbf{x}$.

We shall soon prove that every symmetric game admits a symmetric Nash equilibrium. But first, we consider a biological toy model which played an essential role in the emergence of evolutionary game theory [**27**]. It is due to two eminent theoretical biologists, John Maynard Smith and George Price, who tried to explain the evolution of ritual fighting in animal contests. It had often been observed that in conflicts within a species, animals did not escalate the fight, but kept to certain stereotyped behavior, such as posturing, glaring, roaring or engaging in a pushing match. Signals of surrender (such as offering the unprotected throat) stopped the fight as reliably as a towel thrown into the boxing ring. Interestingly, thus, animal fights seem to be restrained by certain rules, without even needing a referee. Such restraint is obviously all for the good of the species, but Darwinian thinking does not accept this as an argument for its emergence. An animal ignoring these 'gloved fist'-type of rules, and killing its rivals, should be able to spread its genes, and the readiness to escalate a conflict should grow, even if this implies, in the long run, suicide for the species.

Maynard Smith and Price imagined, in their thought experiment, a population consisting of two phenotypes (or strategies). Strategy $\mathbf{e}_1$ is a behavioral program to escalate the conflict until death or injury settles the outcome. Strategy $\mathbf{e}_2$ is a behavioral program to flee as soon as the opponent starts getting rough. The former strategy is called 'Hawk', the latter 'Dove'. Winning the conflict yields an expected payoff G, and losing an escalated fight costs $C > G$ (with G and C measured on the scale of Darwinian fitness). If we assume that whenever two 'Hawks' meet, or two 'Doves', both are equally likely to win the contest, then their expected payoff is $G/2 - C/2$ resp. $G/2$. The payoff matrix thus is

$$\begin{pmatrix} \frac{G-C}{2} & G \\ 0 & \frac{G}{2} \end{pmatrix}. \tag{5.4}$$

Clearly neither 'Hawk' nor 'Dove' is a Nash equilibrium. In terms of evolutionary biology a homogeneous 'Dove' population could be invaded by a minority of 'Hawks' who win all their contests hands down; but similarly, a homogeneous 'Hawk' population could be invaded by a minority of 'Doves', whose payoff 0 is larger than the negative payoff $(G-C)/2$ experienced by the 'Hawks' tearing at each other. It is better to experience no change in reproductive success, rather than a reduction. In this sense neither 'Hawk' nor 'Dove' is an evolutionarily stable strategy. On the basis of this extremely simplified model, we must expect evolution to lead to a mixed population.

6. Population dynamics

Let us consider a symmetric game with payoff matrix A and assume that in a large, well-mixed population, a fraction x_i uses strategy $\mathbf{e}_i$, for $i = 1, ..., n$. The state of the population is thus given by the vector $\mathbf{x} \in \Delta_n$. A player with strategy $\mathbf{e}_i$ has as expected payoff

$$(A\mathbf{x})_i = \sum_j a_{ij} x_j. \tag{6.1}$$

Indeed, this player meets with probability x_j a co-player using $\mathbf{e}_j$. The average payoff in the population is given by

$$\mathbf{x} \cdot A\mathbf{x} = \sum_i x_i (A\mathbf{x})_i. \tag{6.2}$$

It should be stressed that we are committing an abuse of notation. The same symbol $\mathbf{x} \in \Delta_n$ which denoted in the previous sections the mixed strategy of one specific player (cf. (2.1) and (2.2)) now denotes the state of a population consisting of different types, each type playing its pure strategy. (We could also have the players use mixed strategies, but will consider this case only later.)

Now comes an essential step: we shall assume that populations can evolve, in the sense that the relative frequencies x_i change with time. Thus we let the state $\mathbf{x}(t)$ depend on time, and denote by $\dot{x}_i(t)$ the velocity with which x_i changes. The assumption of differentiability implies an infinitely large population, or the interpretation of x_i as an expected value, rather than a bona fide frequency. Both ways of thinking are familiar to mathematical ecologists. In keeping with our population dynamical approach, we shall be particularly interested in the (per capita) growth rates $\dot{x}_i/x_i$ of the frequencies of the strategies.

How do the frequencies of strategies evolve? There are many possibilities for modeling this process. We shall at first assume that the state of the population evolves according to the *replicator equation* (see [**40, 16, 46**] and, for the name, [**37**]). This equation holds if the growth rate of a strategy's frequency corresponds to the strategy's payoff, or more precisely to the difference between its payoff $(A\mathbf{x})_i$ and the average payoff $\mathbf{x} \cdot A\mathbf{x}$ in the population. Thus we posit

$$\dot{x}_i = x_i[(A\mathbf{x}_i) - \mathbf{x} \cdot A\mathbf{x}] \tag{6.3}$$

for $i = 1, ..., n$. Accordingly, a strategy $\mathbf{e}_i$ will spread or dwindle depending on whether it does better or worse than average.

This yields a deterministic model for the state of the population. Before we try to motivate the replicator equation, let us note that $\sum \dot{x}_i{=}0$. Furthermore, it is easy to see that the constant function $x_i(t) = 0$ for all t obviously satisfies the i-th component of equation (6.3). Hence the hyperplanes $\sum x_i = 1$ and $x_i = 0$ are invariant. From this follows that the state space, i.e. the simplex Δ_n, is invariant: if $\mathbf{x}(0) \in \Delta_n$ then $\mathbf{x}(t) \in \Delta_n$ for all $t \in R$. The same holds for all sub-simplices of Δ_n (which are given by $x_i = 0$ for one or several i), and hence also for the boundary $bd\Delta_n$ of Δ_n (i.e. the union of all such sub-simplices), and moreover also for the interior $int\Delta_n$ of the simplex (the subset satisfying $x_i > 0$ for all i). From now on we only consider the restriction of (6.3) to the state simplex Δ_n.

7. Basic properties of the replicator equation

It is easy to see that if we add an arbitrary function $b(\mathbf{x})$ to all payoff terms $(A\mathbf{x})_i$, the replicator equation (6.3) remains unchanged: what is added to the payoff is also added to the average payoff $\mathbf{x} \cdot A\mathbf{x}$, since $\sum x_i = 1$, and cancels out in the difference of the two terms. In particular, this implies that we can add a constant c_j to the j-th column of A (for $j = 1, ..., n$) without altering the replicator dynamics on Δ_n. We shall frequently use this to simplify the analysis.

Another useful property is the quotient rule: if $x_j > 0$, then the time-derivative of the quotient satisfies

$$\left(\frac{x_i}{x_j}\right)^{\cdot} = \left(\frac{x_i}{x_j}\right)[(A\mathbf{x})_i - (A\mathbf{x})_j]. \tag{7.1}$$

Thus the relative proportions of two strategies change according to their payoff ranking. More generally, if $V = \prod x_i^{p_i}$ then

$$\dot{V} = V[\mathbf{p} \cdot A\mathbf{x} - (\sum p_i)\mathbf{x} \cdot A\mathbf{x}]. \tag{7.2}$$

The rest points $\mathbf{z}$ of the replicator equation are those for which all payoff values $(A\mathbf{z})_i$ are equal, for all indices i for which $z_i > 0$. The common value of these payoffs is the average payoff $\mathbf{z} \cdot A\mathbf{z}$. In particular, all vertices $\mathbf{e}_i$ of the simplex Δ_n are rest points. (Obviously, if all players are of the same type, mere copying leads to no change.) The replicator equation admits a rest point in $int\Delta_n$ if there exists a solution (in $int\Delta_n$) of the linear equations

$$(A\mathbf{x})_1 = ... = (A\mathbf{x})_n. \tag{7.3}$$

Similarly, all rest points on each face can be obtained by solving a corresponding system of linear equations. Generically, each sub-simplex (and Δ_n itself) contains one or no rest point in its interior.

One can show that if no rest point exists in the interior of Δ_n, then all orbits in $int\Delta_n$ converge to the boundary, for $t \to \pm\infty$. In particular, if strategy $\mathbf{e}_i$ is *strictly dominated*, i.e., if there exists a $\mathbf{w} \in \Delta_n$ such that $(A\mathbf{x})_i < \mathbf{w} \cdot A\mathbf{x}$ holds for all $\mathbf{x} \in \Delta_n$, then $x_i(t) \to 0$ for $t \to +\infty$ [**21**]. In the converse direction, if there exists an orbit $\mathbf{x}(t)$ bounded away from the boundary of Δ_n (i.e. such that for some $a > 0$ the inequality $x_i(t) > a$ holds for all $t > 0$ and all $i = 1, ..., n$), then there exists a rest point in $int\Delta_n$ [**18**]. One just has to note that for $i = 1, ..., n$,

$$(\log x_i)^{\cdot} = \dot{x}_i/x_i = (A\mathbf{x}(t))_i - \mathbf{x}(t) \cdot A\mathbf{x}(t). \tag{7.4}$$

Integrating for $t \in [0, T]$, and dividing by T, leads on the left hand side to $[\log x_i(T) - \log x_i(0)]/T$, which converges to 0 for $T \to +\infty$. The corresponding limit on the right hand side implies that for the accumulation points z_i of the time averages

$$z_i(T) = \frac{1}{T} \int_0^T x_i(t)dt, \tag{7.5}$$

the relations $z_i \geq a > 0$, $\sum z_i = 1$, and

$$\sum a_{1j} z_j = ... = \sum a_{nj} z_j \tag{7.6}$$

must hold. Using (7.3), we see that $\mathbf{z}$ is a rest point in $int\Delta_n$.

8. The Lotka-Volterra connection

There is an intimate connection between Lotka-Volterra equations, which are the staple fare of mathematical population ecology, and the replicator equation [**18**]. More precisely, there exists a diffeomorphism from $\Delta_n^- = \{\mathbf{x} \in \Delta_n : x_n > 0\}$ onto R_+^{n-1} mapping the orbits of the replicator equation (6.3) onto the orbits of the Lotka-Volterra equation

$$\dot{y}_i = y_i(r_i + \sum_{j=1}^{n-1} d_{ij} y_j), \tag{8.1}$$

where $r_i = a_{in} - a_{nn}$ and $d_{ij} = a_{ij} - a_{nj}$. Indeed, let us define $y_n \equiv 1$ and consider the transformation $\mathbf{y} \to \mathbf{x}$ given by

$$x_i = \frac{y_i}{\sum_{j=1}^n y_j} \qquad i = 1, \ldots, n \tag{8.2}$$

which maps $\{\mathbf{y} \in R_+^n : y_n = 1\}$ onto Δ_n^-. The inverse $\mathbf{x} \to \mathbf{y}$ is given by

$$y_i = \frac{y_i}{y_n} = \frac{x_i}{x_n} \qquad i = 1, \ldots, n\ . \tag{8.3}$$

Now let us consider the replicator equation in n variables given by (6.3). We shall assume that the last row of the $n \times n$ matrix $A = (a_{ij})$ consists of zeros: since we can add constants to columns, this is no restriction of generality. By the quotient rule (7.1)

$$\dot{y}_i = (\frac{x_i}{x_n})[(A\mathbf{x})_i - (A\mathbf{x})_n]. \tag{8.4}$$

Since $(A\mathbf{x})_n = 0$, this implies

$$\dot{y}_i = y_i(\sum_{j=1}^n a_{ij} x_j) = y_i(\sum_{j=1}^n a_{ij} y_j) x_n\ . \tag{8.5}$$

By a change in velocity, we can remove the term $x_n > 0$. Since $y_n = 1$, this yields

$$\dot{y}_i = y_i(a_{in} + \sum_{j=1}^{n-1} a_{ij}y_j) \tag{8.6}$$

or (with $r_i = a_{in}$) equation (8.1).

The converse direction from (8.1) to (6.3) is analogous.

Results about Lotka-Volterra equations can therefore be carried over to the replicator equation and vice versa. Some properties are simpler to prove (or more natural to formulate) for one equation and some for the other.

For instance, it is easy to prove for the Lotka-Volterra equation that the interior of R^n_+ contains α- or ω- limit points if and only if it admits an interior rest point. Indeed, let $L : \mathbf{x} \to \mathbf{y}$ be defined by

$$y_i = r_i + \sum_{j=1}^{n} a_{ij}x_j \qquad i = 1, \ldots, n\ . \tag{8.7}$$

If (8.1) admits no interior rest point, the set $K = L(\text{int } R^n_+)$ is disjoint from $\mathbf{0}$. A well known theorem from convex analysis implies that there exists a hyperplane H through $\mathbf{0}$ which is disjoint from the convex set K. Thus there exists a vector $\mathbf{c}= (c_1, \ldots, c_n) \neq \mathbf{0}$ orthogonal to H (i.e. $\mathbf{c} \cdot \mathbf{x} = 0$ for all $\mathbf{x} \in H$) such that $\mathbf{c} \cdot \mathbf{y}$ is positive for all $\mathbf{y} \in K$. Setting

$$V(\mathbf{x}) = \prod x_i^{c_i}, \tag{8.8}$$

we see that V is defined on int R^n_+. If $\mathbf{x}(t)$ is a solution of (8.1) in int R^n_+, then the time derivative of $t \to V(\mathbf{x}(t))$ satisfies

$$\dot{V} = V \sum c_i \frac{\dot{x}_i}{x_i} = V \sum c_i y_i = V\mathbf{c} \cdot \mathbf{y} > 0\ . \tag{8.9}$$

Thus V is increasing along each orbit. But then no point $\mathbf{z} \in$ int R^n_+ may belong to an ω-limit: indeed, by Lyapunov's theorem, the derivative $\dot{V}$ would have to vanish there. This contradiction completes the proof.

In particular, if $int\Delta_n$ contains a periodic orbit of the replicator equation (6.3), it must also contain a rest point.

9. Two-dimensional examples

Let us discuss the replicator equation when there are only two types in the population. Since the equation remains unchanged if we subtract the diagonal term in each column, we can assume without restricting generality that the 2×2-matrix A is of the form

$$\begin{pmatrix} 0 & a \\ b & 0 \end{pmatrix}. \tag{9.1}$$

Since $x_2 = 1 - x_1$, it is enough to consider x_1, which we denote by x. Thus $x_2 = 1 - x$, and

$$\dot{x} = x[(A\mathbf{x})_1 - \mathbf{x} \cdot A\mathbf{x}] = x[(A\mathbf{x})_1 - (x(A\mathbf{x})_1 + (1-x)(A\mathbf{x})_2)], \tag{9.2}$$

and hence

$$\dot{x} = x(1-x)[(A\mathbf{x})_1 - (A\mathbf{x})_2], \tag{9.3}$$

which reduces to

$$\dot{x} = x(1-x)[a-(a+b)x]. \tag{9.4}$$

We note that

$$a = \lim_{x \to 0} \frac{\dot{x}}{x}. \tag{9.5}$$

Hence a corresponds to the limit of the per capita growth rate of the missing strategy $\mathbf{e}_1$. Let us omit the trivial case $a = b = 0$: in this case all points of the state space Δ_2 (i.e. the interval $0 \leq x \leq 1$) are rest points. The right hand side of our differential equation is a product of three factors, the first vanishing at 0 and the second at 1; the third factor has a zero $\hat{x} = \frac{a}{a+b}$ in $]0,1[$ if and only if $ab > 0$. Thus we obtain three possible cases.

(1) There is no rest point in the interior of the state space. This happens if and only if $ab \leq 0$. In this case, $\dot{x}$ has always the same sign in $]0,1[$. If this sign is positive (i.e. if $a \geq 0$ and $b \leq 0$, at least one inequality being strict), this means that $x(t) \to 1$ for $t \to +\infty$, for every initial value $x(0)$ with $0 < x(0) < 1$. The strategy $\mathbf{e}_1$ is said to *dominate* strategy $\mathbf{e}_2$. It is always the best reply, for any value of $x \in]0,1[$. Conversely, if the sign of $\dot{x}$ is negative, then $x(t) \to 0$ and $\mathbf{e}_2$ dominates. In each case, the dominating strategy converges towards fixation.

As an example, we consider the Prisoner's Dilemma Game from (4.5). The two strategies $\mathbf{e}_1$ and $\mathbf{e}_2$ are usually interpreted as 'cooperation' (by providing a benefit to the co-player) and 'defection' (by refusing to provide a benefit). The payoff matrix is transformed, by adding appropriate constants to each column, into

$$\begin{pmatrix} 0 & -5 \\ 5 & 0 \end{pmatrix} \tag{9.6}$$

and defection dominates.

(2) There exists a rest point $\hat{x}$ in $]0,1[$ (i.e. $ab > 0$), and both a and b are negative. In this case $\dot{x} < 0$ for $x \in]0,\hat{x}[$ and $\dot{x} > 0$ for $x \in]\hat{x},1[$. This means that the orbits lead away from $\hat{x}$: this rest point is unstable. As in the previous case, one strategy will be eliminated: but the outcome, in this *bistable* case, depends on the initial condition. If x is larger than the threshold $\hat{x}$, it will keep growing; if it is smaller, it will vanish – a positive feedback.

As an example, we can consider the coordination game (4.3). The payoff matrix is transformed into

$$\begin{pmatrix} 0 & -2 \\ -2 & 0 \end{pmatrix} \tag{9.7}$$

and it is best to play $\mathbf{e}_1$ if the frequency of $\mathbf{e}_1$-players exceeds 50 percent. Bistability also occurs if the Prisoner's Dilemma game given by (4.5) is repeated sufficiently often. Let us assume that the number of rounds is a random variable with mean value m, for instance, and let us consider only two strategies of particular interest. One, which will be denoted by $\mathbf{e}_1$, is the Tit For Tat strategy which consists in cooperating in the first round and from then on imitating what the co-player did in the previous round. The other strategy, denoted as $\mathbf{e}_2$, consists in always defecting. The expected payoff values are given by the matrix

$$\begin{pmatrix} 10m & -5 \\ 15 & 0 \end{pmatrix} \tag{9.8}$$

which can be transformed into

$$\begin{pmatrix} 0 & -5 \\ 15-10m & 0 \end{pmatrix}. \tag{9.9}$$

If $m > 3/2$, it is best to do what the co-player does. Loosely speaking, one should go with the trend. The outcome, in such a population, would be the establishment of a single norm of behavior (either always defect, or play Tit For Tat). Which norm emerges depends on the initial condition.

(3) There exists a rest point $\hat{x}$ in $]0,1[$ (i.e. $ab > 0$), and both a and b are positive. In this case $\dot{x} > 0$ for $x \in]0,\hat{x}[$ and $\dot{x} < 0$ for $x \in]\hat{x},1[$. This negative feedback means that $x(t)$ converges towards $\hat{x}$, for $t \to +\infty$: the rest point $\hat{x}$ is a stable attractor. No strategy eliminates the other: rather, their frequencies converge towards a *stable coexistence.*

This situation can be found in the Hawk-Dove-game, for example. The payoff matrix (5.4)is transformed into

$$\begin{pmatrix} 0 & \frac{G}{2} \\ \frac{C-G}{2} & 0 \end{pmatrix} \tag{9.10}$$

and the rest point corresponds to $x = G/C$. The higher the cost of injury, i.e., C, the lower the frequency of escalation. Another well-known example is the so-called snowdrift game. Suppose that two players are promised 40 dollars each if they contribute 30 dollars to the experimenter. They have to decide independently whether to come up with such a fee or not. If both contribute, they can split the cost equally, and pay only 15 dollars. If $\mathbf{e}_1$ is the decision to contribute, and $\mathbf{e}_2$ not to contribute, the payoff matrix is

$$\begin{pmatrix} 25 & 10 \\ 40 & 0 \end{pmatrix} \tag{9.11}$$

which can be normalized to

$$\begin{pmatrix} 0 & 10 \\ 15 & 0 \end{pmatrix}. \tag{9.12}$$

In this case, it is best to do the opposite of what the co-player is doing, i.e., to swim against the stream.

10. Rock-Scissors-Paper

Turning now to $n = 3$, we meet a particularly interesting example if the three strategies dominate each other in a cyclic fashion, i.e., if $\mathbf{e}_1$ dominates $\mathbf{e}_2$, in the absence of $\mathbf{e}_3$, and similarly $\mathbf{e}_2$ dominates $\mathbf{e}_3$, and $\mathbf{e}_3$, in turn, dominates $\mathbf{e}_1$. Such a cycle occurs in the game of Rock-Scissors-Paper shown in (1.2). It is a zero-sum game: one player receives what the other player loses. Hence the average payoff in the population, $\mathbf{x} \cdot A\mathbf{x}$, is zero. There exist only four rest points, one in the center, $\mathbf{m} = (1/3, 1/3, 1/3) \in int\Delta_3$, and the other three at the vertices $\mathbf{e}_i$.

Let us consider the function $V := x_1 x_2 x_3$, which is positive in the interior of Δ_3 (with maximum at $\mathbf{m}$) and vanishes on the boundary. Using (7.2), we see that $t \to V(\mathbf{x}(t))$ satisfies

$$\dot{V} = V(x_2 - x_3 + x_3 - x_1 + x_1 - x_2) = 0. \tag{10.1}$$

Hence V is a *constant of motion*: all orbits $t \to \mathbf{x}(t)$ of the replicator equation remain on constant level sets of V. This implies that all orbits in $int\Delta_n$ are closed orbits surrounding $\mathbf{m}$. The invariant set consisting of the three vertices $\mathbf{e}_i$ and the orbits connecting them along the edges of Δ_3 is said to form a *heteroclinic set.* Any two points on it can be connected by 'shadowing the dynamics'. This means to travel along the orbits of that set and, at appropriate times which can be arbitrarily rare, to make an arbitrarily small step. In the present case, it means for instance to flow along an edge from $\mathbf{e}_2$ towards $\mathbf{e}_1$, and then stepping onto the edge leading away from $\mathbf{e}_1$ and toward $\mathbf{e}_3$. This step can be arbitrarily small: travellers just have to wait until they are sufficiently close to the 'junction' $\mathbf{e}_1$.

Now let us consider the generalized Rock-Scissors-Paper game with matrix

$$\begin{pmatrix} 0 & a & -b \\ -b & 0 & a \\ a & -b & 0 \end{pmatrix}. \tag{10.2}$$

with $a, b > 0$, which is no longer zero-sum, if $a \neq b$. It has the same structure of cyclic dominance and the same rest points. The point $\mathbf{m}$ is a Nash equilibrium and the boundary of Δ_3 is a heteroclinic set, as before. But now,

$$\mathbf{x} \cdot A\mathbf{x} = (a - b)(x_1x_2 + x_2x_3 + x_3x_1), \tag{10.3}$$

and hence

$$\dot{V} = V(a - b)[1 - 3(x_1x_2 + x_2x_3 + x_3x_1)], \tag{10.4}$$

which implies

$$\dot{V} = \frac{V(a-b)}{2}[(x_1 - x_2)^2 + (x_2 - x_3)^2 + (x_3 - x_1)^2]. \tag{10.5}$$

This expression vanishes on the boundary of Δ_3 and at $\mathbf{m}$. It has the sign of $a - b$ everywhere else on Δ_3. If $a > b$, this means that all orbits cross the constant-level sets of V in the uphill direction, and hence converge to $\mathbf{m}$. For $a > b$, the function $V(\mathbf{x})$ is a strict *Lyapunov function*: indeed $\dot{V}(\mathbf{x}) \geq 0$ for all $\mathbf{x}$, and equality holds only when $\mathbf{x}$ is a rest point. This implies that ultimately, all three types will be present in the population in equal frequencies: the rest point $\mathbf{m}$ is asymptotically stable. But for $a < b$, the orbits flow downhill towards the boundary of Δ_3. The Nash equilibrium $\mathbf{m}$ corresponds to an unstable rest point, and the heteroclinic cycle on the boundary attracts all other orbits.

Let us follow the state $\mathbf{x}(t)$ of the population, for $a < b$. If the state is very close to a vertex, for instance $\mathbf{e}_1$, it is close to a rest point and hence almost at rest. For a long time, the state does not seem to change. Then, it picks up speed and moves towards the vicinity of the vertex $\mathbf{e}_3$, where it slows down and remains for a much longer time, etc. This looks like a recurrent form of 'punctuated equilibrium': long periods of quasi-rest followed by abrupt upheavals.

The same holds if all the a's and b's, in (10.2), are distinct positive numbers. There exists a unique rest point $\mathbf{m}$ in the interior of Δ_3 which, depending on the sign of $\det A$ (which is the same as that of $\mathbf{m} \cdot A\mathbf{m}$) is either globally stable, i.e., attracts all orbits in $int\Delta_3$, or is surrounded by periodic orbits, or is repelling. In the latter case, all orbits converge to the heteroclinic cycle formed by the boundary of Δ_n.

Interestingly, several biological examples for Rock-Scissors-Paper cycles have been found. We only mention two examples: (A) Among the lizard species *Uta stansburiana*, three inheritable types of male mating behavior are $\mathbf{e}_1$: attach yourself to a female and guard her closely, $\mathbf{e}_2$: attach yourself to several females and guard them (but inevitably, less closely); and $\mathbf{e}_3$: attach yourself to no female, but roam around and attempt sneaky matings whenever you encounter an unguarded female [**39**]. (B) Among the bacteria *E. coli*, three strains occur in the lab through recurrent mutations, namely $\mathbf{e}_1$: the usual, so-called *wild type*; $\mathbf{e}_2$: a mutant producing colicin, a toxic substance, together with a protein conferring auto-immunity; and $\mathbf{e}_3$: a mutant producing the immunity-conferring protein, but not the poison [**23**]. In case (A), selection leads to the stable coexistence of all three types, and in case (B) to the survival of one type only.

There exist about 100 distinct phase portraits of the replicator equation for $n = 3$, up to re-labeling the vertices [**1**]. Of these, about a dozen are generic. Interestingly, none admits a limit cycle [**19**]. For $n > 3$, limit cycles and chaotic attractors can occur. A classification seems presently out of reach.

11. Nash equilibria and saturated rest points

Let us consider a symmetric $n \times n$-game with payoff matrix A, and $\mathbf{z}$ a symmetric Nash equilibrium. With $\mathbf{x} = \mathbf{e}_i$, condition (5.2) implies

$$(A\mathbf{z})_i \leq \mathbf{z} \cdot A\mathbf{z} \tag{11.1}$$

for $i = 1, ..., n$. Equality must hold for all i such that $z_i > 0$. Hence $\mathbf{z}$ is a rest point of the replicator dynamics. Moreover, it is a *saturated* rest point: this means by definition that if $z_i = 0$, then

$$(A\mathbf{z})_i - \mathbf{z} \cdot A\mathbf{z} \leq 0. \tag{11.2}$$

Conversely, every saturated rest point is a Nash equilibrium. The two concepts are equivalent.

Every rest point in $int\Delta_n$ is trivially saturated; but on the boundary, there may be rest points which are not saturated, as we shall presently see. In that case, there exist strategies which are not present in the population $\mathbf{z}$, but which would do better than average (and better, in fact, than every type that is present). Rest points and Nash equilibria have in common that there exists a c such that $(A\mathbf{z})_i = c$ whenever $z_i > 0$; the additional requirement, for a Nash equilibrium, is that $(A\mathbf{z})_i \leq c$ whenever $z_i = 0$.

Hence every symmetric Nash equilibrium is a rest point, but the converse does not hold. Let us discuss this for the examples from the previous section. It is clear that the rest points in the interior of the simplex are Nash equilibria. In case $n = 2$ and dominance, the strategy that is dominant is a Nash equilibrium, and the other is not. In case $n = 2$ with bi-stability, both pure strategies are Nash equilibria. Generically (and in contrast to the example (9.7)), one of the pure strategies fares better than the other in a population where both are equally frequent. This is the so-called risk-dominant equilibrium. It has the larger basin of attraction. In the case $n = 2$ leading to stable co-existence, none of the pure strategies is a Nash equilibrium. If you play a bistable game, you should choose the same strategy as your co-player; but in the case of stable coexistence, you should choose the opposite strategy. In both cases, however, the two of you might have different ideas about who plays what.

In the case $n = 3$ with the Rock-Scissors-Paper structure, the interior rest point $\mathbf{m}$ is the unique Nash equilibrium. Each of the vertex rest points can be invaded.

A handful of results about Nash equilibria and rest points of the replicator dynamics are known as *folk theorem of evolutionary game theory* [**5**]. For instance, any limit, for $t \to +\infty$, of a solution $\mathbf{x}(t)$ starting in $int\Delta_n$ is a Nash equilibrium; and any stable rest point is a Nash equilibrium. (A rest point $\mathbf{z}$ is said to be stable if for any neighborhood U of $\mathbf{z}$ there exists a neighborhood V of $\mathbf{z}$ such that if $\mathbf{x}(0) \in V$ then $\mathbf{x}(t) \in U$ for all $t \geq 0$). Both results are obvious consequences of the fact that if $\mathbf{z}$ is not Nash, then there exists an i and an ϵ such that $(A\mathbf{x})_i - \mathbf{x} \cdot A\mathbf{x} > \epsilon$ for all $\mathbf{x}$ close to $\mathbf{z}$. In the other direction, if $\mathbf{z}$ is a strict Nash equilibrium, then $\mathbf{z}$ is an asymptotically stable rest point (i.e. not only stable, but in addition *attracting* in the sense that for some neighborhood U of $\mathbf{z}$, $\mathbf{x}(0) \in U$ implies $\mathbf{x}(t) \to \mathbf{z}$ for $t \to +\infty$). The converse statements are generally not valid.

In order to prove the existence of a symmetric Nash equilibrium for the symmetric game with $n \times n$ matrix A, i.e. the existence of a saturated rest point for the corresponding replicator equation (6.3), we perturb that equation by adding a small constant term $\epsilon > 0$ to each component of the right hand side. Of course, the relation $\sum \dot{x}_i = 0$ will no longer hold. We compensate this by subtracting the term $n\epsilon$ from each growth rate $(A\mathbf{x})_i - \mathbf{x} \cdot A\mathbf{x}$. Thus we consider

$$\dot{x}_i = x_i[(A\mathbf{x})_i - \mathbf{x} \cdot A\mathbf{x} - n\epsilon] + \epsilon. \tag{11.3}$$

Clearly, $\sum \dot{x}_i = 0$ is satisfied again. On the other hand, if $x_i = 0$, then $\dot{x}_i = \epsilon > 0$. This influx term changes the vector field of the replicator equation: at the boundary of Δ_n (which is invariant for the unperturbed replicator equation), the vector field of the perturbed equation points towards the interior.

Brouwer's fixed point theorem implies that (11.3) admits at least one rest point in $int\Delta_n$, which we denote by $\mathbf{z}_\epsilon$. It satisfies

$$(A\mathbf{z}_\epsilon)_i - \mathbf{z}_\epsilon \cdot A\mathbf{z}_\epsilon = \epsilon(n - \frac{1}{(\mathbf{z}_\epsilon)_i}). \tag{11.4}$$

Let ϵ tend to 0, and let $\mathbf{z}$ be an accumulation point of the $\mathbf{z}_\epsilon$ in Δ_n. The limit on the left hand side exists, and is given by $(A\mathbf{z})_i - \mathbf{z} \cdot A\mathbf{z}$. Hence the right hand side also has a limit for $\epsilon \to 0$. This limit is 0 if $z_i > 0$, and it is ≤ 0 if $z_i = 0$. This implies that $\mathbf{z}$ is a saturated rest point of the (unperturbed) replicator equation (6.3), and hence corresponds to a Nash equilibrium (see also [**15, 38**]).

12. Mixed strategies and evolutionary stability

Let us now consider the case when individuals can also use mixed strategies, for instance escalate a conflict with a certain probability. Thus let us assume that there exist N types, each using a (pure or mixed) strategy $\mathbf{p}(i) \in \Delta_n$ (we need not assume $n = N$). The average payoff for a $\mathbf{p}(i)$-player against a $\mathbf{p}(j)$-player is given by $u_{ij} = \mathbf{p}(i) \cdot A\mathbf{p}(j)$, and if $\mathbf{x} \in \Delta_N$ describes the frequencies of the types in the population, then the average strategy within the population is $\mathbf{p}(\mathbf{x}) = \sum x_i \mathbf{p}(i)$. The induced replicator dynamics on Δ_N, namely $\dot{x}_i = x_i[(U\mathbf{x})_i - \mathbf{x} \cdot U\mathbf{x}]$ can be written as

$$\dot{x}_i = x_i[(\mathbf{p}(i) - \mathbf{p}(\mathbf{x})) \cdot A\mathbf{p}(\mathbf{x})]. \tag{12.1}$$

This dynamics on Δ_N induces a dynamics $t \to \mathbf{p}(\mathbf{x}(t))$ of the average strategy on Δ_n.

Let us now turn to the concept of an evolutionarily stable strategy, or ESS. If all members of the population use such a strategy $\hat{\mathbf{p}} \in \Delta_n$, then no mutant minority using another strategy $\mathbf{p}$ can invade (cf. [**29, 25**]). Thus a strategy $\hat{\mathbf{p}} \in \Delta_n$ is said to be *evolutionarily stable* if for every $\mathbf{p} \in \Delta_n$ with $\mathbf{p} \neq \hat{\mathbf{p}}$, the induced replicator equation describing the dynamics of the population consisting of these two types only (the 'resident' using $\hat{\mathbf{p}}$ and the 'invader' using $\mathbf{p}$) leads to the elimination of the invader. By (9.4) this equation reads (if x is the frequency of the invader):

$$\dot{x} = x(1-x)[x(\mathbf{p}\cdot A\mathbf{p} - \hat{\mathbf{p}}\cdot A\mathbf{p}) - (1-x)(\hat{\mathbf{p}}\cdot A\hat{\mathbf{p}} - \mathbf{p}\cdot A\hat{\mathbf{p}})] \tag{12.2}$$

and hence the rest point $x = 0$ is asymptotically stable iff the following conditions are satisfied:

(a) (equilibrium condition)

$$\mathbf{p}\cdot A\hat{\mathbf{p}} \leq \hat{\mathbf{p}}\cdot A\hat{\mathbf{p}} \tag{12.3}$$

holds for all $\mathbf{p} \in \Delta_n$;

(b) (stability condition)

$$\text{if}\quad \mathbf{p}\cdot A\hat{\mathbf{p}} = \hat{\mathbf{p}}\cdot A\hat{\mathbf{p}} \quad\text{then}\quad \mathbf{p}\cdot A\mathbf{p} < \hat{\mathbf{p}}\cdot A\mathbf{p}. \tag{12.4}$$

The first condition means that $\hat{\mathbf{p}}$ is a Nash equilibrium: no invader does better than the resident, against the resident. The second condition states that if the invader does as well as the resident against the resident, then it does less well than the resident against the invader. Based on (7.2), it can be shown that the strategy $\hat{\mathbf{p}}$ is an ESS iff $\prod_i x_i^{\hat{p}_i}$ is a strict local Lyapunov function for the replicator equation, or equivalently iff

$$\hat{\mathbf{p}}\cdot A\mathbf{p} > \mathbf{p}\cdot A\mathbf{p} \tag{12.5}$$

for all $\mathbf{p} \neq \hat{\mathbf{p}}$ in some neighborhood of $\hat{\mathbf{p}}$ [**16, 18**]. If $\hat{\mathbf{p}} \in int\Delta_n$, then Δ_n itself is such a neighborhood.

In particular, an ESS corresponds to an asymptotically stable rest point of (6.3). The converse does not hold in general [**46**]. But the strategy $\hat{\mathbf{p}} \in \Delta_n$ is an ESS iff it is *strongly stable* in the following sense: whenever it belongs to the convex hull of $\mathbf{p}(1), ..., \mathbf{p}(N) \in \Delta_n$, the strategy $\mathbf{p}(\mathbf{x}(t))$ converges to $\hat{\mathbf{p}}$, under (12.1), for all $\mathbf{x} \in \Delta_N$ for which $\mathbf{p}(\mathbf{x})$ is sufficiently close to $\hat{\mathbf{p}}$ [**4**].

The relation between evolutionary and dynamic stability is particularly simple for the class of *partnership games*. These are defined by payoff matrices $A = A^T$. In this case the interests of both players coincide. For spartnership games, $\hat{\mathbf{p}}$ is an ESS iff it is asymptotically stable for (6.3). This in turn holds iff it is a strict local maximum of the average payoff $\mathbf{x}\cdot A\mathbf{x}$ [**18**]. Replicator equations for partnership games occur prominently in population genetics. They describe the effect of selection on the frequencies x_i of alleles i on a single genetic locus, for $i \in \{1, ..., n\}$. In this case, the a_{ij} correspond to the survival probabilities of individuals with genotype (i, j) (i.e., having inherited the alleles i and j from their parents).

13. Generalizations of the replicator dynamics

We have assumed so far that the average payoff for a player using strategy i is given by a linear function $(A\mathbf{x})_i$ of the state of the population. This makes sense if the interactions are pairwise, with co-players chosen randomly within the population. But many interesting examples lead to non-linear payoff functions $a_i(\mathbf{x})$, for instance if the interactions occur in groups with more than two members. This is the case, for instance, in the sex-ratio game, where the success of a strategy (i.e., an individual sex ratio) depends on the aggregate sex ratio in the population. Nonlinear payoff functions $a_i(\mathbf{x})$ lead to the replicator equation

$$\dot{x}_i = x_i(a_i(\mathbf{x}) - \bar{a}) \tag{13.1}$$

on Δ_n, where $\bar{a} = \sum_i x_i a_i(\mathbf{x})$ is again the average payoff within the population. Many of the previous results can be extended in a straightforward way. For instance, the dynamics is unchanged under addition of a function b to all payoff functions a_i. Equation (13.1) always admits a saturated rest point, and a straight extension of the folk theorem is still valid. The notion of an ESS has to be replaced by a localized version.

Initially, the replicator dynamics was intended to model the transmission of behavioral programs through inheritance. The simplest inheritance mechanisms lead in a straightforward way to (6.3), but more complex cases of Mendelian inheritance through one or several genetic loci yield more complex dynamics [**13, 7, 45, 17**]. The replicator equation (6.3) can also be used to model imitation processes [**14, 2, 36, 34**]. A rather general approach to modeling imitation processes leads to

$$\dot{x}_i = x_i[f(a_i(\mathbf{x})) - \sum x_j f(a_j(\mathbf{x}))] \tag{13.2}$$

for some strictly increasing function f of the payoff, and even more generally to the *imitation dynamics* given by

$$\dot{x}_i = x_i g_i(\mathbf{x}) \tag{13.3}$$

where the functions g_i satisfy $\sum x_i g_i(\mathbf{x}) = 0$ on Δ_n. The simplex Δ_n and its faces are invariant. Such an equation is said to be *payoff monotonic* if

$$g_i(\mathbf{x}) > g_j(\mathbf{x}) \Leftrightarrow a_i(\mathbf{x}) > a_j(\mathbf{x}), \tag{13.4}$$

where the a_i correspond to the payoff for strategy i. For payoff monotonic equations (13.3), the *folk theorem* holds again [**31, 8**]: Nash equilibria are rest points, strict Nash equilibria are asymptotically stable, and rest points that are stable or ω-limits of interior orbits are Nash equilibria.

The dynamics (13.3) can be reduced (through a change in velocity) to a replicator equation (13.1) if it has the following property:

$$\mathbf{y} \cdot \mathbf{g}(\mathbf{x}) > \mathbf{z} \cdot \mathbf{g}(\mathbf{x}) \Longleftrightarrow \mathbf{y} \cdot \mathbf{a}(\mathbf{x}) > \mathbf{z} \cdot \mathbf{a}(\mathbf{x}) \tag{13.5}$$

for all $\mathbf{x}, \mathbf{y}, \mathbf{z} \in \Delta_n$.

14. Best reply dynamics

It is worth emphasizing that imitation (like selection, in genetics) does not produce anything new. If a strategy $\mathbf{e}_i$ is absent from the population, it will remain so (i.e. if $x_i(t) = 0$ holds for some time t, it holds for all t). An equation such as (13.1) or more generally (13.3) does not allow the introduction of new strategies. There exist game dynamics which are more innovative. For instance, clever players could adopt the strategy which offers the highest payoff, even if no one in the population is currently using it. We describe this dynamics presently. Other innovative dynamics arise if we assume a steady rate of switching randomly to other strategies. This can be interpreted as an 'exploration rate', and corresponds to a mutation term in genetics [**35**].

The best-reply dynamics assumes more sophistication than mere learning by copying others. Let us assume that in a large population, a small fraction of the players revise their strategy, choosing best replies $BR(\mathbf{x})$ to the current mean population strategy $\mathbf{x}$. This approach, which postulates that players are intelligent enough to know the current population state and to respond optimally, yields the *best reply dynamics*

$$\dot{\mathbf{x}} \in BR(\mathbf{x}) - \mathbf{x}. \tag{14.1}$$

Since best replies are in general not unique, this is a differential *inclusion* rather than a differential equation [**26**]. For continuous payoff functions $a_i(\mathbf{x})$, the set of best replies $BR(\mathbf{x})$ is a non-empty convex, compact subset of Δ_n which is upper semi-continuous in $\mathbf{x}$. Hence solutions exist, they are Lipschitz functions $\mathbf{x}(t)$ satisfying (14.1) for almost all $t \geq 0$. If $BR(\mathbf{x})$ is a uniquely defined (and hence pure) strategy $\mathbf{b}$, the solution of (14.1) is given by

$$\mathbf{x}(t) = (1 - e^{-t})\mathbf{b} + e^{-t}\mathbf{x} \tag{14.2}$$

for small $t \geq 0$, which describes a linear orbit pointing straight towards the best response. This can lead to a state where $\mathbf{b}$ is no longer the unique best reply. But for each $\mathbf{x}$ there always exists a $\mathbf{b} \in BR(\mathbf{x})$ which, among all best replies to $\mathbf{x}$, is a best reply against itself (i.e. a Nash equilibrium of the game restricted to the simplex $BR(\mathbf{x})$) [**20**]. In this case $\mathbf{b} \in BR((1-\epsilon)\mathbf{x} + \epsilon\mathbf{b})$ holds for small $\epsilon \geq 0$, if the game is linear. An iteration of this construction yields at least one piecewise linear solution of (14.1) starting at $\mathbf{x}$ and defined for all $t > 0$. One can show that for generic linear games, essentially all solutions can be constructed in this way. For the resulting (multi-valued) semi-dynamical system, the simplex Δ_n is only forward invariant and $bd\Delta_n$ need no longer be invariant: the frequency of strategies which are initially missing can grow, in contrast to the imitation dynamics. In this sense, the best reply dynamics is an *innovative* dynamics.

For $n = 2$, the phase portraits of (14.1) differ only in details from that of the replicator dynamics. If $\mathbf{e}_1$ is dominated by $\mathbf{e}_2$, there are only two orbits: the rest point $\mathbf{e}_2$, and the semi-orbit through $\mathbf{e}_1$ which converges to $\mathbf{e}_2$. In the bistable situation with interior Nash equilibrium $\mathbf{p}$, there are infinitely many solutions starting at $\mathbf{p}$ besides the constant one, staying there for some time and then converging monotonically to either $\mathbf{e}_1$ or $\mathbf{e}_2$. In the case of stable coexistence with interior Nash equilibrium $\mathbf{p}$, the solution starting at some point $\mathbf{x}$ between $\mathbf{p}$ and $\mathbf{e}_1$ converges toward $\mathbf{e}_2$ until it hits $\mathbf{p}$, in finite time, and then remains there forever.

For $n = 3$, the differences to the replicator dynamics become more pronounced. In particular, for the generalized Rock-Scissors-Paper game given by (10.2), all orbits converge to the Nash equilibrium $\mathbf{p}$ whenever $\det A > 0$ (just as with the replicator dynamics); but for $\det A < 0$, all orbits (except possibly $\mathbf{p}$) converge to a limit cycle, the so-called *Shapley triangle* spanned by the three points A_i (given by the intersections of the lines $(A\mathbf{x})_2 = (A\mathbf{x})_3$ etc. in Δ_3). In fact, the piecewise linear function $V(\mathbf{x}) := |\max_i (A\mathbf{x})_i|$ is a Lyapunov function for (14.1). In this case, the orbits of the replicator equation (6.3) converge to the boundary of Δ_n; but interestingly, the time averages

$$\mathbf{z}(T) := \frac{1}{T}\int_0^T \mathbf{x}(t)dt \tag{14.3}$$

have the Shapley triangle as the set of accumulation points, for $T \to +\infty$. Similar parallels between the best reply dynamics and the behavior of time-averages of the replicator equation are quite frequent [**9, 10**].

15. A brief look at asymmetric games

So far, we have considered evolutionary games in the symmetric case only. Thus players are indistinguishable (except by their strategies), and the game is described by a single $n \times n$ payoff matrix A. In the first section, however, we had started out with two players I and II having strategies $\mathbf{e}_i$ and $\mathbf{f}_j$ respectively (with $1 \leq i \leq n$ and $1 \leq j \leq m$), and a game was defined by two $n \times m$ payoff matrices A and B There is an obvious way to turn the non-symmetric game (A, B) into a symmetric game: simply by letting a coin toss decide who of the two players will be labeled player I. A strategy for this *symmetrized* game must therefore specify what to do in role I, and what in role II, i.e., such a strategy is given by a pair $(\mathbf{e}_i, \mathbf{f}_j)$. A mixed strategy is given by an element $\mathbf{z} = (z_{ij}) \in \Delta_{nm}$, where z_{ij} denotes the probability to play $\mathbf{e}_i$ when in role I and $\mathbf{f}_j$ when in role II. To the probability distribution $\mathbf{z}$ correspond its *marginals*: $x_i = \sum_j z_{ij}$ and $y_j = \sum_i z_{ij}$. The vectors $\mathbf{x} = (x_i)$ and $\mathbf{y} = (y_j)$ belong to Δ_n and Δ_m, respectively.

The expected payoff for a player using $(\mathbf{e}_i, \mathbf{f}_j)$ against a player using $(\mathbf{e}_k, \mathbf{f}_l)$, with $i, k \in \{1, ..., n\}$ and $j, l \in \{1, ..., m\}$, is given by

$$c_{ij,kl} = \frac{1}{2}a_{il} + \frac{1}{2}b_{kj}. \tag{15.1}$$

Since every symmetric game has a symmetric Nash equilibrium, it follows immediately that every game (A, B) has a Nash equilibrium pair.

Let us now turn to population games. Players meet randomly and engage in a game (A, B), with chance deciding who is in role I and who in role II. For simplicity, we assume that there are only two strategies for each role. The payoff matrix is

$$\begin{pmatrix} (A, a) & (B, b) \\ (C, c) & (D, d) \end{pmatrix}. \tag{15.2}$$

The strategies for the resulting symmetric game will be denoted by $\mathbf{G}_1 = \mathbf{e}_1\mathbf{f}_1$, $\mathbf{G}_2 = \mathbf{e}_2\mathbf{f}_1$, $\mathbf{G}_3 = \mathbf{e}_2\mathbf{f}_2$ and $\mathbf{G}_4 = \mathbf{e}_1\mathbf{f}_2$. The payoff for a player using $\mathbf{G}_i$ against a player using $\mathbf{G}_j$ is given, up to the factor 1/2 which we shall henceforth omit, by the (i, j)-entry of the matrix

$$M = \begin{pmatrix} A+a & A+c & B+c & B+a \\ C+a & C+c & D+c & D+a \\ C+b & C+d & D+d & D+b \\ A+b & A+d & B+d & B+b \end{pmatrix}. \tag{15.3}$$

This corresponds to (15.1). For instance, a $\mathbf{G}_1$-player meeting a $\mathbf{G}_3$-opponent is with probability 1/2 in role I, plays $\mathbf{e}_1$ against the co-player's $\mathbf{f}_2$, and obtains B. With probability 1/2, the $\mathbf{G}_1$-player is in role II, plays $\mathbf{f}_1$ against the co-players' $\mathbf{e}_2$, and obtains c.

The replicator dynamics

$$\dot{x}_i = x_i[(M\mathbf{x})_i - \mathbf{x} \cdot M\mathbf{x}] \tag{15.4}$$

describes the evolution of the state $\mathbf{x} = (x_1, x_2, x_3, x_4) \in \Delta_4$. Since the dynamics is unaffected if each m_{ij} is replaced by $m_{ij} - m_{1j}$ (for $i, j \in \{1, 2, 3, 4\}$), we can use the matrix

$$\begin{pmatrix} 0 & 0 & 0 & 0 \\ R & R & S & S \\ R+r & R+s & S+s & S+r \\ r & s & s & r \end{pmatrix} \tag{15.5}$$

with $R := C - A$, $r := b - a$, $S := D - B$ and $s := d - c$. We shall denote this matrix again by M. It has the property that

$$m_{1j} + m_{3j} = m_{2j} + m_{4j} \tag{15.6}$$

for $j = 1, 2, 3, 4$. Hence

$$(M\mathbf{x})_1 + (M\mathbf{x})_3 = (M\mathbf{x})_2 + (M\mathbf{x})_4 \tag{15.7}$$

holds for all $\mathbf{x}$. From this and (7.2) follows that the function $V = x_1x_3/x_2x_4$ satisfies

$$\dot{V} = V[(M\mathbf{x})_1 + (M\mathbf{x})_3 - (M\mathbf{x})_2 - (M\mathbf{x})_4] = 0 \tag{15.8}$$

in the interior of Δ_4, and hence that V is an invariant of motion for the replicator dynamics: its value remains unchanged along every orbit.

Therefore, the interior of the state simplex Δ_4 is foliated by the surfaces

$$W_K := \{\mathbf{x} \in \Delta_4 : x_1x_3 = Kx_2x_4\}, \tag{15.9}$$

with $0 < K < \infty$. These are saddle-like surfaces which are spanned by the quadrangle of edges $\mathbf{G}_1\mathbf{G}_2$, $\mathbf{G}_2\mathbf{G}_3$, $\mathbf{G}_3\mathbf{G}_4$ and $\mathbf{G}_4\mathbf{G}_1$ joining the vertices of the simplex Δ_4.

The orientation of the flow on the edges can easily be obtained from the previous matrix. For instance, if $R = 0$, then the edge $\mathbf{G}_1\mathbf{G}_2$ consists of rest points. If $R > 0$, the flow along the edge points from $\mathbf{G}_1$ towards $\mathbf{G}_2$ (which means that in the absence of the strategies $\mathbf{G}_3$ and $\mathbf{G}_4$, the strategy $\mathbf{G}_2$ dominates $\mathbf{G}_1$), and conversely, if $R < 0$, the flow points from $\mathbf{G}_2$ to $\mathbf{G}_1$.

Generically, the parameters R, S, r and s are non-zero. This corresponds to 16 orientations of the quadrangle $\mathbf{G}_1\mathbf{G}_2\mathbf{G}_3\mathbf{G}_4$, which by symmetry can be reduced to 4. Since $(M\mathbf{x})_1$ trivially vanishes, the rest points in the interior of the simplex Δ_4 must satisfy $(M\mathbf{x})_i = 0$ for $i = 2, 3, 4$. This implies for $S \neq R$

$$x_1 + x_2 = \frac{S}{S - R}, \tag{15.10}$$

and for $s \neq r$

$$x_1 + x_4 = \frac{s}{s - r}. \tag{15.11}$$

Such solutions lie in the simplex if and only if $RS < 0$ and $rs < 0$. If this is the case, one obtains a line of rest points which intersects each W_K in exactly one point. These points can be written as

$$x_i = m_i + \xi \tag{15.12}$$

for $i = 1, 3$ and

$$x_i = m_i - \xi \tag{15.13}$$

for $i = 2, 4$, with ξ as parameter and

$$\mathbf{m} = \frac{1}{(S - R)(s - r)}(Ss, -Sr, Rr, -Rs) \in W_1. \tag{15.14}$$

Of particular interest is the so-called Wright-manifold W_1, where the strategies, in the two roles, are independent of each other. (On W_1, the probability that a randomly chosen individual uses strategy $\mathbf{e}_1\mathbf{f}_1$ is the product of the probabilities $x := x_1 + x_4$ and $y := x_1 + x_2$ of choosing $\mathbf{e}_1$ when in role I, resp. $\mathbf{f}_1$ when in role II. Indeed, $x_1 = (x_1 + x_4)(x_1 + x_2)$). It then follows that

$$\dot{x} = x(1-x)(s-(s-r)y), \tag{15.15}$$

and

$$\dot{y} = y(1-y)(S-(S-R)x). \tag{15.16}$$

If $rR > 0$, each interior rest point is a saddle point within the corresponding manifold W_K, and the system is bistable: depending of the inital condition, orbits converge either to G_1 or to G_3, if $r < 0$, and either to G_2 or to G_4, if $r > 0$. If $rR < 0$, each rest point has (in addition of the eigenvalue 1) a pair of complex conjugate eigenvalues. Within the corresponding manifold W_K, the eigenvalues spiral around this rest point. Depending on whether K is larger or smaller than 1, they either converge to the rest point (which must be a spiral sink), or else toward the heteroclinic cycle defined by the quadrangle of the edges forming the boundary of W_K. For $K = 1$, the orbits are periodic.

16. Applications

In this lecture course, the authors aim to stress the variety of plausible dynamics which describe adaptive mechanisms underlying game theory. The replicator equation and the best reply dynamics describe just two out of many dynamics. For applications of evolutionary game theory, it does not suffice to specify the strategies and the payoff values. One also has to be explicit about the transmission mechanisms describing how strategies spread within a population.

We end this introductory part with some signposts to the literature using evolutionary games to model specific social interactions. The first applications, and indeed the motivation, of evolutionary game theory are found in evolutionary biology, where by now thousands of papers have proved the fruitfulness of this approach, see [**6**]. In fact, questions of sex-ratio, and more generally of sex-allocation, even pre-date any explicit formulation in terms of evolutionary game theory. It was R.F. Fisher, a pioneer in both population genetics and mathematical statistics, who used frequency-dependent selection to explain the prevalence of a 1:1 sex ratio, and W.D. Hamilton who extended this type of thinking to make sense of other, odd sex ratios [**12**]. We have seen how Price and Maynard Smith coined their concept of evolutionary stability to explain the prevalence of ritual fighting in intraspecific animal contests. The subtleties of such contests are still a favorite topic among the students of animal behavior. More muted, but certainly not less widespread conflicts arise on the issues of mate choice, parental investment, and parent-offspring conflicts. Social foraging is another field where the success of a given behavior (scrounging, for instance) depends on its prevalence; so are dispersal and habitat selection. Communication (alarm calls, threat displays, sexual advertisement, gossip), with all its opportunities for deceit, is replete with game theoretical problems concerning bluff and honest signaling. Predators and their prey, or parasites and their hosts, offer examples of games between two populations, with the success of a trait depending on the state of the other population. Some strategic interactions are surprisingly sophisticated, considering the lowly level of the players: for instance, bacteria can engage in quorum sensing as cue for conditional behavior.

Quite a few biological games turned out to have the same structure as games that had been studied by economists, usually under another name [**3**]: the biologists' 'Hawk-Dove' game, for example, has the same structure as the economists' 'Chicken'-game. Evolutionary game theory has found a large number of applications in economic interactions [**44, 22, 41, 8, 11**].

One zone of convergence for studies of animal behavior and human societies is that of cooperation. Indeed, the theory of evolution and economic theory have each their own paradigm of selfishness, encapsulated in the slogans of the 'selfish gene' and the 'homo economicus'. Both paradigms conflict with wide-spread evidence of social, 'other-regarding' behavior. In ant and bee societies, the relatedness of individuals is so close that their genetic interests overlap and their communities can be viewed as 'super-organisms'. But in human societies, close cooperation can also occur between individuals who are unrelated. In many cases, such cooperation is based on reciprocation. Positive and negative incentives, and in particular the threat of sanctions offer additional reasons for the prevalence of cooperation [**38**]. This may lead to two or more stable equilibria, corresponding to behavioral norms. If everyone adopts a given norm, no player has an incentive to deviate. But which of these norms eventually emerges depends, among other things, on the history of the population.

Animal behavior and experimental economics fuse in this area. Experimental economics, has greatly flourished in the last few years. It often reduces to the investigation of very simple games which can be analyzed by means of evolutionary dynamics. These and other games display the limitations of 'rational' behavior in humans, and have assisted in the emergence of new fields, such as behavioral game theory and neuro-economics.

References

1. I.M. Bomze, *Non-cooperative two-person games in biology: a classification*, Int. J. Game Theory **15** (1986), 31-57.
2. T. Börgers and R. Sarin, *Learning through reinforcement and replicator dynamics*, J. Economic Theory **77** (1997), 1-14.
3. A.M. Colman, Game Theory and its Applications in the Social and Biological Sciences, Oxford: Butterworth-Heinemann (1995).
4. R. Cressman, The Stability concept of Evolutionary Game Theory, Springer, Berlin (1992).
5. R. Cressman, Evolutionary Dynamics and Extensive Form Games, MIT Press (2003)
6. L.A. Dugatkin and H.K. Reeve (eds.), Game Theory and Animal Behavior, Oxford UP (1998).
7. I. Eshel, *Evolutionarily stable strategies and viability selection in Mendelian populations*, Theor. Population Biology **22** (1982), 204-217.
8. D. Friedman, *Evolutionary games in economics*, Econometrica **59** (1991), 637-66.
9. A. Gaunersdorfer, *Time averages for heteroclinic attractors*, SIAM J. Appl. Math **52** (1992), 1476-89.
10. A. Gaunersdorfer and J. Hofbauer, *Fictitious play, Shapley polygons and the replicator equation*, Games and Economic Behavior **11** (1995), 279-303.
11. H. Gintis, Game Theory Evolving, Princeton UP (2000).
12. W.D. Hamilton, *Extraordinary sex ratios*, Science **156**, 477-488.
13. P. Hammerstein and R. Selten, *Game theory and evolutionary biology*, in R.J. Aumann, S. Hart (eds.), Handbook of Game Theory II, Amsterdam, North-Holland (1994), 931-993.
14. D. Helbing, *Interrelations between stochastic equations for systems with pair interactions*, Physica A **181** (1992), 29-52.

15. J. Hofbauer, *From Nash and Brown to Maynard Smith: equilibria, dynamics and ESS*, Selection **1** (2000), 81-88.
16. J. Hofbauer, P. Schuster, and K. Sigmund, *A note on evolutionarily stable strategies and game dynamics*, J. Theor. Biology **81** (1979), 609-612.
17. J. Hofbauer, P. Schuster, and K. Sigmund, *Game dynamics for Mendelian populations*, Biol. Cybernetics **43** (1982), 51-57.
18. J. Hofbauer and K. Sigmund, The Theory of Evolution and Dynamical Systems, Cambridge UP (1988).
19. J. Hofbauer and K. Sigmund, *Evolutionary Games and Population Dynamics*, Cambridge UP (1998).
20. J. Hofbauer and K. Sigmund, *Evolutionary game dynamics*, Bulletin of the American Mathematical Society **40**, (2003) 479-519.
21. J. Hofbauer and J. W. Weibull, *Evolutionary selection against dominated strategies*, J. Economic Theory **71** (1996), 558-573.
22. M. Kandori: *Evolutionary Game Theory in Economics*, in D. M. Kreps and K. F. Wallis (eds.), Advances in Economics and Econometrics: Theory and Applications, I, Cambridge UP (1997).
23. B. Kerr, M.A. Riley, M.W. Feldman, and B.J.M. Bohannan, *Local dispersal promotes biodiversity in a real-life game of rock-paper-scissors*, Nature **418** (2002), 171-174.
24. R. Leonard, Von Neumann, Morgenstern and the Creation of Game Theory: from Chess to Social Science, 1900-1960, Cambridge, Cambridge UP (2010).
25. S. Lessard, *Evolutionary stability: one concept, several meanings*, Theor. Population Biology **37** (1990), 159-70.
26. A. Matsui, *Best response dynamics and socially stable strategies*, J. Econ. Theory **57** (1992), 343-362.
27. J. Maynard Smith and G. Price, *The logic of animal conflict*, Nature **246** (1973), 15-18.
28. J. Maynard Smith, *Will a sexual population converge to an ESS?*, American Naturalist **177** (1981), 1015-1018.
29. J. Maynard Smith, Evolution and the Theory of Games, Cambridge UP (1982)
30. R. Myerson, Game Theory: Analysis of Conflict, Cambridge, Mass., Harvard University Press (1997)
31. J. Nachbar, *"Evolutionary" selection dynamics in games: convergence and limit properties*, Int. J. Game Theory **19** (1990), 59-89.
32. S. Nasar, A Beautiful Mind: A Biography of John Forbes Nash, Jr., Winner of the Nobel Prize in Economics, New York, Simon and Schuster (1994).
33. J. Nash, *Non-cooperative games*, Ann. Math. **54** (1951), 287-295.
34. M.A. Nowak, Evolutionary Dynamics, Cambridge MA, Harvard UP (2006).
35. W.H. Sandholm, Population Games and Evolutionary Dynamics, Cambridge, MA, MIT Press (2010).
36. K.H Schlag, *Why imitate, and if so, how? A boundedly rational approach to multi-armed bandits*, J. Econ. Theory **78** (1997), 130-156.
37. P. Schuster and K. Sigmund, *Replicator Dynamics*, J. Theor. Biology **100** (1983), 533-538.
38. K. Sigmund, The Calculus of Selfishness, Princeton, Princeton UP (2010).
39. B. Sinervo and C.M. Lively, *The rock-paper-scissors game and the evolution of alternative male strategies*, Nature **380** (1996), 240-243.
40. P.D. Taylor and L. Jonker, *Evolutionarily stable strategies and game dynamics*, Math. Biosciences **40** (1978), 145-156.
41. F. Vega-Redondo, Evolution, Games, and Economic Theory, Oxford UP (1996).
42. J. von Neumann, *Zur Theorie der Gesellschaftsspiele*, Mathematische Annalen **100** (1928), 295-320.
43. J. von Neumann and O. Morgenstern *Theory of Games and Economic Behavior*, Princeton UP (1944).
44. J. Weibull, Evolutionary Game Dynamics, MIT Press, Cambridge, Mass. (1995).
45. F. Weissing: *Evolutionary stability and dynamic stability in a class of evolutionary normal form games*, in R. Selten (ed.) *Game Equilibrium Models I*, Berlin, Springer (1991), 29-97.
46. E.C. Zeeman, *Population dynamics from game theory*, in *Global Theory of Dynamical Systems*, Springer Lecture Notes in Mathematics **819** (1980).

Faculty of Mathematics, University of Vienna, A-1090 Vienna, Austria and International Institute for Applied Systems Analysis, A-2361 Laxenburg, Austria
E-mail address: karl.sigmund@univie.ac.at

Proceedings of Symposia in Applied Mathematics
Volume **69**, 2011

Beyond the Symmetric Normal Form: Extensive Form Games, Asymmetric Games and Games with Continuous Strategy Spaces

Ross Cressman

Abstract. Evolutionary games are typically introduced through developing theory and applications for symmetric normal form games. This chapter generalizes evolutionary game theory to three other classes of games that are equally important; namely, extensive form games, asymmetric games and games with continuous strategy spaces. Static solution concepts such as Nash equilibrium (NE) and evolutionarily stable strategy (ESS) are extended to these games and connections made with the stability of deterministic evolutionary dynamics (specifically, the replicator equation). The similarities as well as the differences with the corresponding concepts from symmetric normal form games are highlighted. The theory is illustrated through numerous well-known examples from the literature.

Introduction

The initial development of evolutionary game theory and evolutionary stability typically assumed
1. that an individual's payoff depends either on his strategy and that of his opponent used during a single interaction with another player (normal form game) or on his strategy and the current behavioral distribution of the population through a single random interaction (population game; playing-the-field model),
2. that the (pure) strategy set S available to an individual is finite and the same for each player (symmetric game).
In this chapter these assumptions are relaxed in three different ways and the consequences are investigated for evolutionary dynamics, especially the replicator equation.

First, suppose that pairs of individuals have a series of interactions with each other and that the set of actions available at later interactions may depend on what choices were made in earlier ones. Many parlour games (e.g. tic-tac-toe, chess) are

2000 *Mathematics Subject Classification.* Primary 91A22.
The author thanks Sabin Lessard for his thorough review and constructive comments. Also appreciated is the assistance of Francisco Franchetti and Bill Sandholm in preparing the phase diagrams for the figures showing trajectories of the replicator equation. The software is available at W. H. Sandholm, E. Dokumaci, and F. Franchetti (2011). Dynamo: Diagrams for Evolutionary Game Dynamics, version 1.0. http://www.ssc.wisc.edu/ whs/dynamo.

of this sort and it is my contention that most important "real-life" games involving humans or other species include a series of interactions among the same individuals. It is often more appropriate to represent such games in extensive form rather than normal form (Section 1).

Second, in many cases, it is more reasonable to assume that strategies available to one player are different than those available to another. For instance, choices available when Black moves in chess are not usually the same as for White. Similarly, if players are from two different species, their strategy sets will almost surely be different (e.g. predator and prey). Suppose that there are two (or more) types of players and a finite set of strategies for each type. If there are exactly two types and the only interactions are single ones between a player of each type, we have a bimatrix game. Otherwise, it is a more general asymmetric game in either extensive or normal form (Section 2).

Finally, Section 3 considers briefly symmetric (asymmetric) games where the pure strategy set for each (type of) player is a continuum such as a subinterval of real numbers. Now the replicator equation is an infinite dimensional dynamical system on the space(s) of probability measures over the subinterval(s) that correspond to the distribution(s) of individual behaviors. Generalizations of the ESS (evolutionarily stable strategy) concept can be defined that characterize stability of single strategies (i.e. Dirac delta distributions) under the replicator equation as well as under the simpler canonical equation of adaptive dynamics that approximates the evolution of the mean distribution(s).

In these three sections, each standard result taken from the literature is given as a Theorem, with a reference where its proof can be found. Partial proofs are provided here for some of the Theorems when they complement the presentation in the main text.

1. Extensive Form Games

Although (finite, two-player) extensive form games are most helpful when used to represent a game with long (but finite) series of interactions between the same two players, differences with normal form intuition already emerge for short games with perfect information.

1.1. Perfect information games. A (finite, two-player) *perfect information game* is given by a rooted game tree Γ where each non-terminal node is a decision point of one of the players or of nature. A path to a node x is a sequence of edges and nodes connecting the root to x. The edges leading away from the root at each player decision node are this player's choices (or actions) at this node. There must be at least two choices at each player decision node and any such choice that does not yield a terminal node must be on some path to a decision node of the other player. A pure (behavior) strategy for a player specifies a choice at all of his decision nodes. A mixed behavior strategy for a player specifies a probability distribution over the set of actions at each of his decision nodes. Payoffs to both players are specified at each terminal node $z \in Z$. A probability distribution over Z is called an outcome.

Example 1. (Weibull, 1995) Figure 1 is an elementary perfect information game with no moves by nature (i.e. at each non-terminal node, either player 1 or player 2 has a decision point). At each terminal node, payoffs to both players are indicated

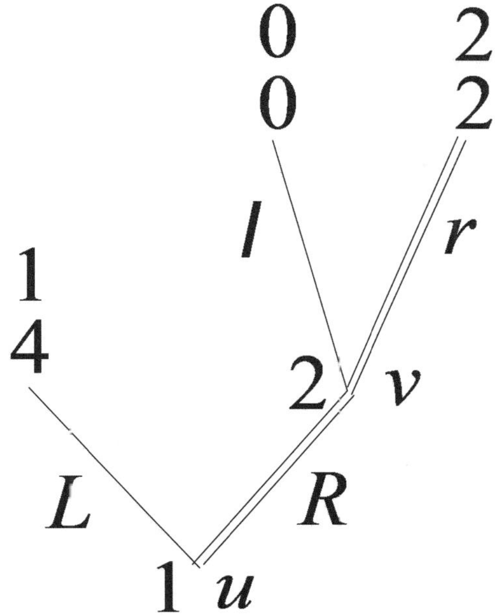

FIGURE 1. Extensive form for Example 1.

with the payoff of player 1 above that of player 2. Player 1 has one decision node u where he chooses between the actions L and R. If he takes action L, player 1 gets payoff 1 and player 2 gets 4. If he takes action R, then we reach the decision point v of player 2 who then chooses between ℓ and r leading to both players receiving payoff 0 or both payoff 2 respectively.

What are the Nash equilibria (NE) for this example? If players 1 and 2 choose R and r respectively with payoff 2 for both, then
1. player 2 does worse through unilaterally changing his strategy by playing r with probability q less than 1 (since $0(1-q)+2q<2$) and
2. player 1 does worse through unilaterally changing his strategy by playing L with positive probability p (since $1p+2(1-p)<2$).
Thus, the strategy pair (R,r) is a strict NE corresponding to the outcome $(2,2)$.[1]

In fact, if player 1 plays R with positive probability at a NE, then player 2 must play r. From this it follows that player 1 must play R with certainty (i.e. $p=0$) (since his payoff of 2 is better than 1 obtained by switching to L). Thus any NE with $p<1$ must be (R,r). On the other hand, if $p=1$ (i.e. player 1 chooses L), then player 2 is indifferent to what strategy he uses since his payoff is 4 for any (mixed) behavior. Furthermore, player 1 is no better off by playing R with positive probability if and only if player 2 plays ℓ at least half the time (i.e. $0\le q\le\frac{1}{2}$). Thus

$$G\equiv\{(L,(1-q)\ell+qr\mid 0\le q\le\frac{1}{2}\}$$

[1]Recall that a NE is strict if each player does worse by unilaterally changing his strategy. When the outcome is a single node, this is understood by saying the outcome is the payoff pair at this node.

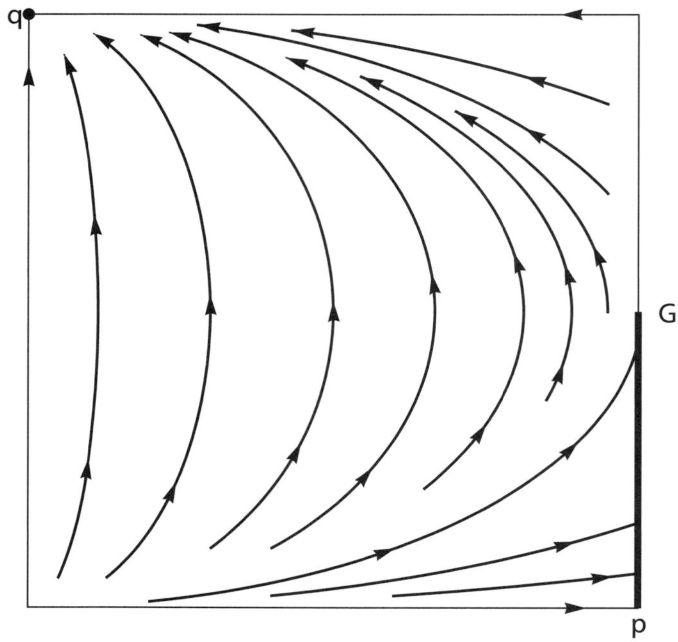

FIGURE 2. Trajectories of the replicator equation for Example 1.

is a set of NE, all corresponding to the outcome $(1,4)$. G is called a NE component since it is a connected set of NE that is not contained in any larger connected set of NE. The NE structure of Example 1 consists of the single strategy pair $G^* = \{(R,r)\}$ and the set G. These are indicated as a solid point and line segment respectively in Figure 2 where $G^* = \{(p,q) \mid p = 0, q = 1\} = \{(0,1)\}$.

Remark 1. Example 1 is a famous game known as the Entry Deterrence Game or the Chain Store Game introduced by the Nobel laureate Reinhard Selten (Selten, 1978). Player 2 is a monopolist who wants to keep the potential entrant (player 1) from entering the market that has a total value of 4. He does this by threatening to ruin the market (play ℓ with both payoffs 0) if player 1 enters (plays R), rather than accepting the entrant (play r and split the total value of 4 to yield payoff 2 for each player). However, this is often viewed as an incredible (i.e. unbelievable) threat since the monopolist should accept the entrant if his decision point is reached (i.e. if player 1 enters) since this gives the higher payoff to him (i.e. $2 > 0$).

Some game theorists argue that a generic perfect information game has only one rational NE equilibrium outcome and this can be found by backward induction. This procedure starts at a final player decision point (i.e. a player decision point that has no player decision points following it) and decides which unique action this player chooses there to maximize his payoff in the subgame with this as its root. The original game tree is then truncated at this node by creating a terminal node there with payoffs to the two players given by this action. The process is continued until the game tree has no player decision nodes left and yields the subgame perfect NE (SPNE). That is, the strategy constructed by backward induction produces a NE in each subgame Γ_u corresponding to the subtree with root at the decision node

u (Kuhn, 1953). For generic perfect information games (see Remark 2), the SPNE is a unique pure strategy pair and is indicated by the double lines in the game tree. If a NE is not subgame perfect, then this perspective argues that there is some player decision node where an incredible threat would be used.

Example 1. (Continued) Can evolutionary dynamics be used to select one of the two NE outcomes of the Chain Store Game? Suppose players 1 and 2 use mixed strategies p and q respectively. The payoffs of pure strategies L and R are 1 and $(1-q)0+2q$ respectively and the payoffs of pure strategies ℓ and r are $4p+(1-p)0$ and $4p+(1-p)2$ respectively. Thus, the expected payoffs are $p+(1-p)2q$ and $(1-q)4p+q(4p+(1-p)2)$ for players 1 and 2 respectively. Under the replicator equation, the probability of using a pure strategy increases if its payoff is higher than these expected payoffs. For this example, the replicator equation is (Weibull, 1995)

$$
\begin{aligned}
(1.1)\ \dot{p} &= p(1-(p+(1-p)2q)) = p(1-p)(1-2q) \\
\dot{q} &= q(4p+(1-p)2-[(1-q)4p+q(4p+(1-p)2)]) = q(1-q)2(1-p).
\end{aligned}
$$

The rest points are the two vertices $\{(0,0),(0,1)\}$ and the edge $\{(1,q) \mid 0 \le q \le 1\}$ joining the other two vertices. Notice that, for any interior trajectory, q is strictly increasing and that p is strictly increasing (decreasing) if and only $q < \frac{1}{2}$ $(q > \frac{1}{2})$. Trajectories of (1.1) are shown in Figure 2.

The following results for Example 1 are straightforward to prove.
1. Every NE outcome is a single terminal node.[2]
2. Every NE component G includes a pure strategy pair.
3. The outcomes of all elements of G are the same.
4. Every interior trajectory of the replicator equation converges to a NE.
5. Every pure strategy NE is stable but not necessarily asymptotically stable.
6. Every NE that has a neighborhood whose only rest points are NE is stable.
7. If a NE component is interior attracting, it includes the SPNE.
8. Suppose (p,q) is a NE. It is asymptotically stable if and only if it is strict. Furthermore, (p,q) is asymptotically stable if and only if playing this strategy pair reaches every player decision point with positive probability (i.e. (p,q) is pervasive).

From Result 8, the SPNE of the Chain Store Game is the only asymptotically stable NE. That is, asymptotic atability of the evolutionary dynamics selects a unique outcome for Example 1 whereby player 1 enters the market and the monopolist is forced to accept this. In general, we have the following theorem.

Theorem 1. (Cressman, 2003) Results 2 to 8 are true for all generic perfect information games. Result 1 holds for generic perfect information games without moves by nature.

Remark 2. By definition, an extensive form game Γ is *generic* if no two pure strategy pairs that yield different outcomes have the same payoff for one of the players. If Γ is a perfect information game and there are no moves by nature, this is equivalent to the property that no two terminal nodes have the same payoff for one of the players. If Γ is not generic, the SPNE outcome may not be unique since

[2]For Example 1, this is either $(2,2)$ or $(1,4)$.

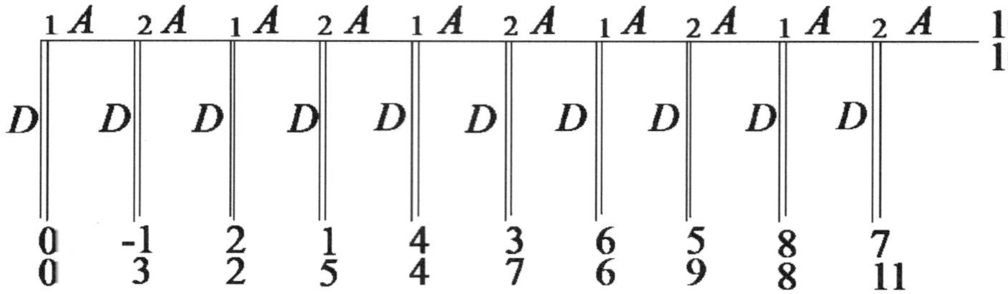

FIGURE 3. Centipede game of length ten.

several choices may arise at some player decision point in the backward induction process if there are payoff ties. Some of the results of Theorem 1 are true for general perfect information games and some are not. For instance, Result 1 is not true for some non-generic games or for generic games with moves by nature. Result 4, which provides the basis to connect dynamics with NE in Results 5 to 8, remains an open problem for non-generic perfect information games.

Theorem 1 applies to all generic perfect information games such as those in Figures 3 and 4. For the centipede game of Figure 3 (Rosenthal, 1981), the SPNE is for both players to play D (down) at each of their five decision points. In fact, the only NE outcome is $(0, 0)$ (i.e. player 1 plays D immediately) and so every interior trajectory converges to a NE with this outcome (the dynamics here is in a $2(2^5 - 1) = 62$ dimensional space). Note that, at each player decision point besides the last, both players are better off if this player plays A (across) there and his opponent plays A at the next decision point. From this it follows that, if any choice D is eliminated, then the SPNE of the new perfect information game is the terminal node that immediately follows the last D eliminated. Centipede games of any fixed length can be easily constructed with the same properties as Figure 3. They provide a large class of perfect information games for which no NE is asymptotically stable since the only equilibrium outcome is not pervasive (cf. Theorem 1, Results 8). These games with a long chain of decision points are often used to question the rationality assumptions behind equilibrium behavior when such outcomes have many unreached decision points.

Since no pure strategy pair in Figure 4 can reach both the left-side subgame and the right-side subgame, none are pervasive. Thus no NE can be asymptotically stable by Theorem 1 (Results 1 and 8), a more elementary example than Figure 3. In fact, Figure 4 is probably the easiest example (Cressman, 2003) of a perfect information game where the NE component G^* of the SPNE outcome $(2, 3)$ is not interior attracting (i.e. there are interior initial points arbitrarily close to G^* whose interior trajectory under the replicator equation does not converge to this NE component). That is, Figure 4 illustrates that the converse of Result 7 (Theorem 1) is not true.

To see this, some notation is needed. The (mixed) strategy space of player 1 is the one-dimensional strategy simplex $\Delta(\{T, B\}) = \{(p_T, p_B) \mid p_T + p_B = 1, 0 \leq p_T, p_B \leq 1\}$. This is also denoted $\Delta^2 \equiv \{(p_1, p_2) \mid p_1 + p_2 = 1, 0 \leq$

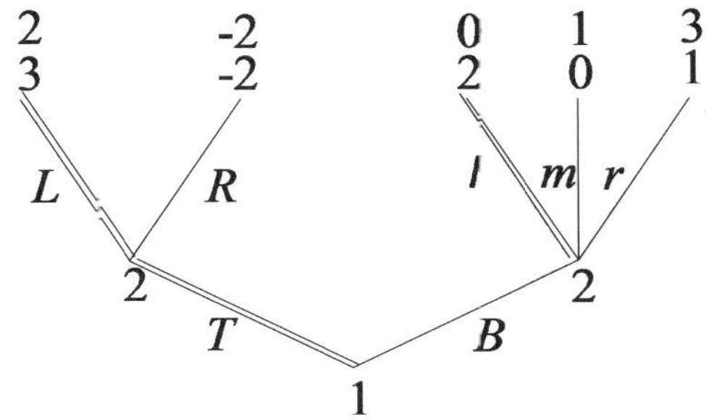

FIGURE 4. Perfect information game with unstable SPNE component.

$p_i \leq 1\}$.[3] Similarly, the strategy simplex for player 2 is the five-dimensional set $\Delta(\{L\ell, Lm, Lr, R\ell, Rm, Rr\}) = \{(q_{L\ell}, q_{Lm}, q_{Lr}, q_{R\ell}, q_{Rm}, q_{Rr}) \in \Delta^6\}$. The replicator equation is then a dynamics on the 6 dimensional space $\Delta(\{T,B\}) \times \Delta(\{L\ell, Lm, Lr, R\ell, Rm, Rr\})$. The SPNE component (i.e. the NE component containing the SPNE) is

$$G^* = \{(T,q) \mid q_{L\ell} + q_{Lm} + q_{Lr} = 1, q_{Lm} + 3q_{Lr} \leq 2\}$$

corresponding to the set of strategy pairs with outcome $(2,3)$ where neither player can improve his payoff by unilaterally changing his strategy. For example, if player 1 switches to B, his payoff of 2 changes to $0q_{L\ell} + 1q_{Lm} + 3q_{Lr} \leq 2$. The only other pure strategy NE is $\{B, R\ell\}$ with outcome $(0,2)$ and corresponding NE component $G = \{(B,q) \mid q_{L\ell} + q_{R\ell} = 1, \frac{1}{2} \leq q_{R\ell} \leq 1\}$. In particular, $(T, \frac{1}{2}q_{Lm} + \frac{1}{2}q_{Lr}) \in G^*$ and $(B, R\ell) \in G$.

The face $\Delta(\{T,B\}) \times \Delta(\{Lm, Lr\})$ has the same structure as the Chain Store Game of Example 1. Specifically, the dynamics is given in Figure 2 where p corresponds to the probability player 1 uses T and q the probability player 2 uses Lr. Thus, points in the interior of this face with $q_{Lr} > \frac{1}{2}$ that start close to $(T, \frac{1}{2}q_{Lm} + \frac{1}{2}q_{Lr})$ converge to (B, Lr). The weak domination of Lm by Lr implies $q_{Lr}(t) > q_{Lr}(t)$ for $t \geq 0$ along all trajectories starting sufficiently close to these points. Since the payoff to B is larger than to T if $q_{Lr}(t) > q_{Lr}(t)$ on the face $\Delta(\{T,B\}) \times \Delta(\{Lm, Lr, R\ell\})$, it can be shown that such trajectories on this face converge to the NE (B, Lr), which is a strict NE for the game restricted to this strategy space. By the stability of (B, Lr) for the full game (Theorem 1, Result 5) and the continuous dependence of trajectories over finite time intervals on initial conditions, there are trajectories in the interior of the full game that start arbitrarily close to G^* that converge to a point in the NE component of G. That is, G^* is not interior attracting.

The partial dynamic analysis of Figure 4 given in the preceding two paragraphs illustrates nicely how the extensive form structure (i.e. the game tree for this perfect information game) helps with properties of NE and the replicator equation (see also Remark 6).

[3]In general, Δ^n is the set of vectors in $\mathbf{R}^n$ with nonnegative components that sum to 1.

Remark 3. Extensive form games can always be represented in normal form. The bimatrix normal form of Example 1 is

$$\begin{array}{r} \\ \text{Not Enter } (L) \\ \text{Enter } (R) \end{array} \begin{array}{c} \text{Ruin } (\ell) \quad \text{Accept } (r) \\ \left[\begin{array}{cc} 1,4 & 1,4 \\ 0,0 & 2,2 \end{array}\right] \end{array}.$$

By convention, player 1 is the row player and player 2 the column player. Each bimatrix entry specifies payoffs received (with player 1's given first) when the two players use their corresponding pure strategy pair. The bimatrix normal forms are also denoted $\left[A, B^T\right]$ where A and B are the payoff matrices for player 1 and 2 respectively. In Example 1,

$$A = \left[\begin{array}{cc} 1 & 1 \\ 0 & 2 \end{array}\right] \text{ and } B = \left[\begin{array}{cc} 4 & 4 \\ 0 & 2 \end{array}\right]^T = \left[\begin{array}{cc} 4 & 0 \\ 4 & 2 \end{array}\right].$$

This elementary example already shows a common feature of the normal form approach for such games; namely, that some payoff entries are repeated in the bimatrix. As a normal form, this means the game is non-generic even though it arose from a generic perfect information game. For this reason, most normal form games cannot be represented as perfect information games. However, they can all be represented as one-stage simultaneity games (e.g. Figure 7).

1.2. Simultaneity games. A (finite, two-player) *simultaneity game* is an extensive form game that involves n stages such that, at each stage, both players know all actions that have already occurred in previous stages but not the opponent's action in the current stage. If there are moves by nature, each of these nodes occur at the beginning of some stage and both players know what action nature takes at these nodes. Thus, the first player decision point x at any stage is the root of a subgame (and so an information set u for this player). Any decision node y following x in the same stage must be a decision point of the other player. The set v of all such y forms an information set for this other player and so each y in v has the same set of actions. A choice by this player at information set v is given by taking the same action at each of the decision points y in v. The simultaneity game is symmetric if there is a bijection between the information sets and actions of player 1 and those of player 2 that makes its normal form representation a symmetric game (as explained in the following example based on Figure 5).

For what follows, it is important to understand both the standard normal form and the reduced-strategy normal form of an extensive form game with game tree Γ. A player's pure strategy for the standard normal form of Γ specifies a choice at each of his information sets. On the other hand, a player's pure strategy for the reduced-strategy normal form specifies a choice at only those information sets v of this player which are relevant given choices already specified by this strategy at information sets of this player on the path to v. Here, v is relevant if there exists a pure strategy of the other player so that v is reached if this pure-strategy pair is used. In Section 1.1, the standard normal form is identical to the reduced-strategy normal form for Figure 1 and for Figure 4. However, for the centipede game of Figure 3, there are $2^5 = 32$ pure strategies for player 1 for the standard normal form (specifying a choice between Across and Down at each of his five decision points) whereas there are only 6 pure strategies $(D, AD, AAD, AAAD, AAAAD, AAAAA)$ in the reduced-strategy normal form. The replicator equation of Section 1.1 is

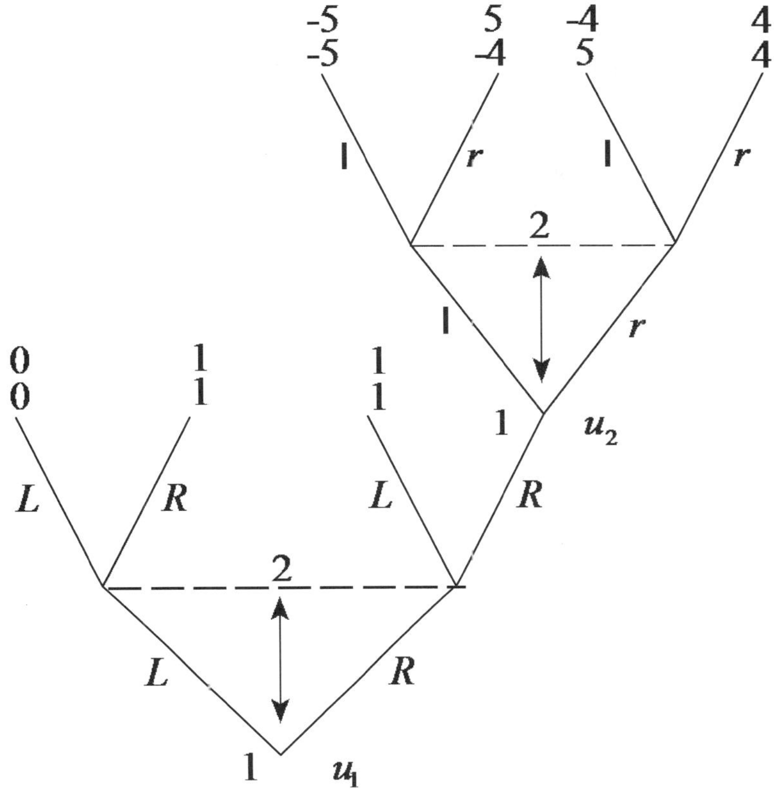

FIGURE 5. Extensive form of Example 2.

implicitly based on the standard normal form, although the results of Theorem 1 (when suitably interpreted) remain true for the reduced-strategy normal form.

Example 2. (van Damme, 1991) Figure 5 is an example of an elementary two-stage symmetric simultaneity game. The two information sets of player 2 both include two decision points and are indicated by dashed horizontal lines. At each of these information sets, player 2 must make the same choice at both decision points. The reduced-strategy normal form is

$$\begin{array}{c} L \\ R\ell \\ Rr \end{array} \begin{array}{c} \begin{array}{ccc} L & R\ell & Rr \end{array} \\ \left[\begin{array}{ccc} 0,0 & 1,1 & 1,1 \\ 1,1 & -5,-5 & 5,-4 \\ 1,1 & -4,5 & 4,4 \end{array}\right] \end{array}.$$

This is a symmetric game since the column player (player 2) has payoff matrix $\begin{bmatrix} 0 & 1 & 1 \\ 1 & -5 & -4 \\ 1 & 5 & 4 \end{bmatrix}^T$ which is the same as the payoff matrix $A = \begin{bmatrix} 0 & 1 & 1 \\ 1 & -5 & 5 \\ 1 & -4 & 4 \end{bmatrix}$ of the row player (player 1). Thus, a player's payoff depends only on the strategy pair used and not on his designation as a row or column player.

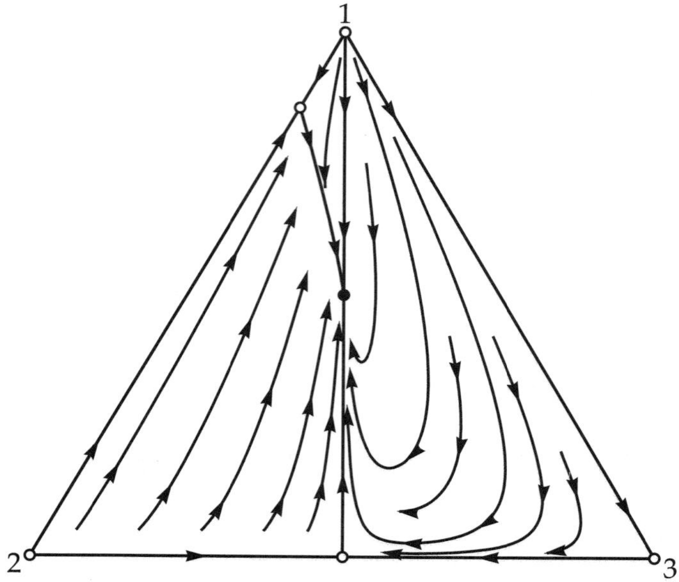

FIGURE 6. Trajectories of the replicator equation for Example 2.

To apply backward induction to this example, the only proper subgame Γ_{u_2} has root at u_2 and payoff matrix given by[4]

$$A_{u_2} = \begin{matrix} \ell \\ r \end{matrix} \begin{bmatrix} -5 & 5 \\ -4 & 4 \end{bmatrix}.$$

This is equivalent to a Hawk-Dove Game with a unique symmetric NE $\frac{1}{2}\ell + \frac{1}{2}r$ (which is also an ESS) and corresponding payoff 0. The truncated game with root at u_1 has payoff matrix

$$\begin{matrix} L \\ R \end{matrix} \begin{bmatrix} 0 & 1 \\ 1 & 0 \end{bmatrix}$$

and it also has unique ESS at $\frac{1}{2}L + \frac{1}{2}R$ and corresponding payoff $\frac{1}{2}$. Thus, $\frac{1}{2}L + \frac{1}{2}\left(\frac{1}{2}R\ell + \frac{1}{2}Rr\right) = \frac{1}{2}L + \frac{1}{4}R\ell + \frac{1}{4}Rr$ is a symmetric NE of Example 2 which can be easily confirmed since

$$Ap^* = \begin{bmatrix} 1/2 \\ 1/2 \\ 1/2 \end{bmatrix} \text{ where } p^* = \begin{bmatrix} 1/2 \\ 1/4 \\ 1/4 \end{bmatrix}.$$

Somewhat surprisingly, p^* is not an ESS of A since, for example, $e_3 \cdot Ap^* = p^* \cdot Ap^* = \frac{1}{2}$ and $p^* \cdot Ae_3 < e_3 \cdot Ae_3$ (i.e. $\frac{1}{2} + \frac{5}{4} + \frac{4}{4} < 4$).[5] However, p^* is globally asymptotically stable under the replicator equation (Figure 6) by the following theorem.

Recall that the replicator equation for a symmetric game with $n \times n$ payoff matrix A is

$$\dot{p}_i = (e_i - p) \cdot Ap$$

[4] Γ_{u_2} is a symmetric subgame and so only the payoffs to player 1 are required in A_{u_2}.

[5] Recall that p^* is an ESS of a symmetric normal form game with $n \times n$ payoff matrix A if i) $p \cdot Ap^* \leq p^* \cdot Ap^*$ for all $p \in \Delta^n$ and ii) $p^* \cdot Ap > p \cdot Ap$ whenever $p \cdot Ap^* = p^* \cdot Ap^*$ and $p \neq p^*$.

for $i = 1, ..., n$ where e_i is the unit vector (corresponding with the ith pure strategy) that has 1 in its ith component and 0 everywhere else and p_i is the proportion of the population using strategy e_i.

Theorem 2. (Cressman, 2003) Suppose that Γ is a symmetric simultaneity game. If p^* is an asymptotically stable NE of the standard normal form of Γ under the replicator equation, then p^* is pervasive and subgame perfect. If Γ has no moves by nature, then a pervasive NE p^* of the reduced-strategy normal form of Γ is asymptotically stable under the replicator equation if and only if p^* is given by backward induction applied to the asymptotically stable pervasive NE of the subgames of Γ and their truncations.

PROOF. First, consider the standard normal form of Γ. For a symmetric extensive form game, a strategy is pervasive if if reaches every player information set when played against itself. If p^* is a NE that is not pervasive, then there is an information set u that is not reached when both players use p^*. Since Γ is a symmetric simultaneity game, we may assume that u is an information set of player 1. Since there are at least two actions at u, we can change p^* to a different strategy p so that p^* and p induce the same behavior strategy at each player 1 information set reachable by p^*. It can be shown that any convex combination of p^* and p is then a rest point of the replicator equation. In particular, none of these points in this connected set can be asymptotically stable. On the other hand, any NE induces a NE in each subgame that it reaches when played against itself (Kuhn, 1953; Selten, 1983). Thus, if p^* is pervasive, it induces a NE in every subgame and so is a SPNE.

Now consider the reduced-strategy normal form of Γ. The proof of the last statement of the theorem is considerably more difficult (see Cressman (2003) for more details). The key is that the replicator equation at p on Γ induces the replicator equation in each subgame Γ_u up to constant multiples given by functions of p that are positive for p close to p^* when p^* is pervasive. Asymptotic stability of p^* is then equivalent to asymptotic stability for the truncated game. The proof is completed by applying these results to a subgame at the last stage of Γ_u and using induction on the number of subgames of Γ. □

Remark 4. Selten mistakenly asserted in 1983 (correcting himself in 1988) that the backward induction procedure applied to the ESSs of subgames and their truncations yields a direct ESS (i.e. an ESS in behavior strategies) for symmetric extensive form games. Example 2 is the well-known counterexample due to Eric van Damme (1991). By Theorem 2, Selten's assertion is true when "ESS" is replaced by "asymptotic stability under the replicator equation" and there are no moves by nature. One must be careful extending this result when there are moves by nature as illustrated by Example 3 below.

Every (symmetric) normal form game can be represented as a single-stage (symmetric) simultaneity game. Thus, unlike perfect information games, generic symmetric simultaneity games can have a NE outcome (and ESS) whose NE component

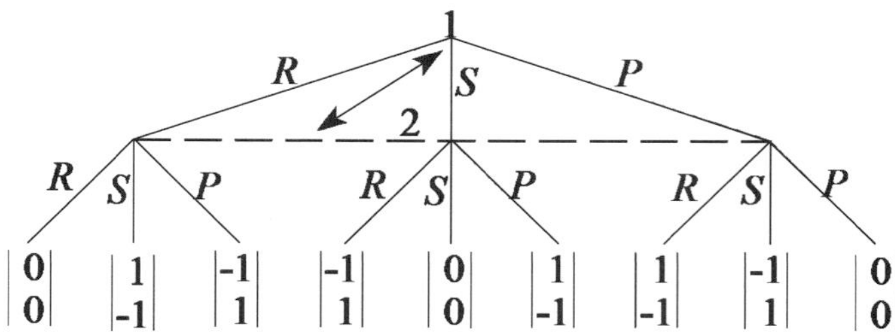

FIGURE 7. Extensive form of the standard RSP Game.

does not include a pure strategy. For instance, the standard zero-sum Rock-Scissors-Paper (RSP) Game with payoff matrix

$$\begin{matrix} R \\ S \\ P \end{matrix} \begin{bmatrix} 0 & 1 & -1 \\ -1 & 0 & 1 \\ 1 & -1 & 0 \end{bmatrix}$$

and unique NE $(1/3, 1/3, 1/3)$ has extensive form given in Figure 7.

The following example uses the generalized RSP game with payoff matrix

$$\begin{bmatrix} 0 & b_2 & -a_3 \\ -a_1 & 0 & b_3 \\ b_1 & -a_2 & 0 \end{bmatrix} = \begin{bmatrix} 0 & 6 & -4 \\ -4 & 0 & 4 \\ 2 & -2 & 0 \end{bmatrix} \tag{1.2}$$

and unique NE $p^* = (10/29, 8/29, 11/29)$. All such games with positive parameters a_i and b_i exhibit cyclic dominance whereby R beats S (i.e. R strictly dominates S in the two-strategy game based on these two strategies), S beats P, and P beats R. They all have a unique NE that is in the interior of Δ^3. From Hofbauer and Sigmund (1998, Section 7.7), p^* is not an ESS for (1.2) since $b_1 < a_3$ (i.e. $2 < 4$) but it is globally asymptotically stable under the replicator equation (Figure 8) since $a_1 a_2 a_3 < b_1 b_2 b_3$ (i.e. $2 \cdot 4 \cdot 4 < 2 \cdot 4 \cdot 6$).

Example 3. (Chamberland and Cressman, 2000) Suppose that, on even numbered days, players play the left-hand subgame of Figure 9 and on odd numbered days the right-hand subgame (an alternative interpretation is that nature flips a fair coin at the root of Figure 9 that determines which subgame is played). However, for both types of days, the same RSP game given by (1.2) is played. In this single-stage symmetric simultaneity game, both players have 9 pure (behavior) strategies $RR, RS, ..., PP$ that specify a choice of R, S or P in each of the subgames. The unique symmetric NE outcome is for both players to play $p^* = (10/29, 8/29, 11/29)$ in both subgames. The corresponding NE component is a four-dimensional set E of points $p = (p_{11}, ..., p_{33})$ formed by intersecting a four-dimensional hyperplane with the eight-dimensional strategy simplex Δ^9. These are the points whose induced marginal strategies in the two subgames (i.e. p^1 and p^2 where $p_i^1 \equiv \sum_{j=1}^3 p_{ij}$ and $p_j^2 \equiv \sum_{i=1}^3 p_{ij}$) are both equal to p^*. It can be shown (Chamberland and Cressman, 2000) that some points in E near the boundary of Δ^9 are unstable since the linearization of the replicator equation there yields an eigenvalue with positive real part.

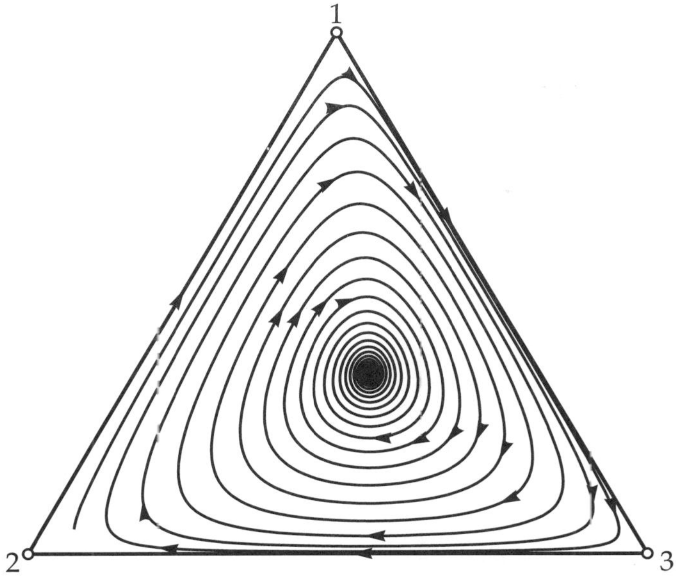

FIGURE 8. Trajectories of the replicator equation for the game with payoff matrix (1.2).

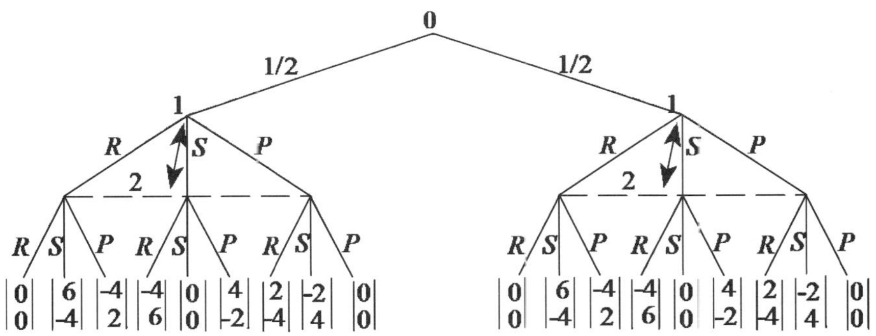

FIGURE 9. Extensive form of a single-stage simultaneity game with a move by nature and identical generalized RSP subgames.

The reason this can occur is that, at a general $p \in \Delta^9$, the evolution of strategy frequencies in one subgame can be influenced by payoffs received in the other subgame. In particular, the frequency of R use in the left-hand subgame can be increasing even if the population state there is mostly P users. To avoid this type of unintuitive situation, the replicator equation can be restricted to the four-dimensional invariant Wright manifold

$$W \equiv \{p \in \Delta^9 \mid p_{ij} = p_i^1 p_j^2\}.$$

On W, the dynamics for the induced strategy in each subgame is the same as the replicator equation for the payoff matrix (1.2). Thus, each interior trajectory that starts on W converges to the single point p^* with $p_{ij}^* = p_i^* p_j^*$.

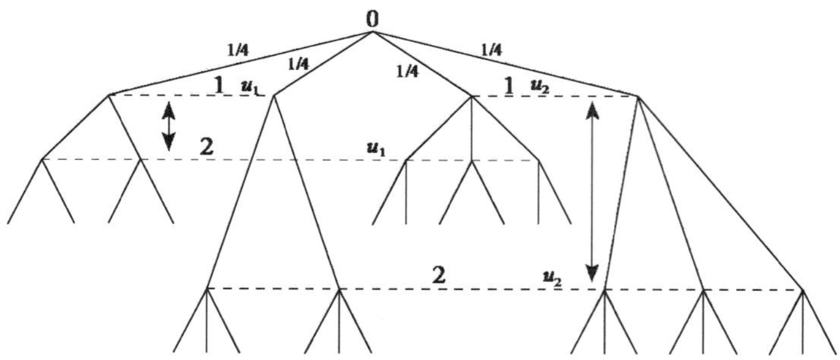

FIGURE 10. Extensive form of an asymmetric two-role game.

Remark 5. The Wright manifold W can be defined for all simultaneity games (in fact, all extensive form games) and it is invariant under the replicator equation. On W, Theorem 2 is true for all symmetric simultaneity games whether or not there are moves by nature.

Remark 6. Extensive form games, with an explicit description of the sequential feature of the players' possible actions, played a central role in the initial development of classical game theory by von Neumann and Morgenstern (1944). On the other hand, most dynamic analyses of evolutionary games are based on their normal forms One consequence of this is that typical normal form examples considered in evolutionary game theory have a small number of pure strategies since it is well-known that the high-dimensional systems of evolutionary dynamics associated to a large number of pure strategies can exhibit all the complexities of arbitrary dynamical systems such as periodic orbits, limit cycles, bifurcations and chaos. The above discussion was meant to convince you that the extensive form structure (which is usually associated with a large number of pure strategies) imparts special properties on the evolutionary dynamics that makes its analysis more tractable than would otherwise be expected.

2. Asymmetric Games

A (finite, two-player) *asymmetric game* has a set $\{u_1, u_2, ..., u_N\}$ of N roles. Players 1 and 2 are assigned roles u_k and u_ℓ respectively with probability $\rho(u_k, u_\ell)$. We assume that role assignment is independent of player designation (i.e. $\rho(u_k, u_\ell) = \rho(u_\ell, u_k)$). If players are assigned the same role (i.e. $k = \ell$), then they play a symmetric (normal form) game with payoff matrix A_{kk}. When they are assigned different roles (i.e. $k \neq \ell$), they play a bimatrix (normal form) game with payoff matrices $A_{k\ell}$ and $A_{\ell k}$.

Figure 10 is the extensive form of a two role game with two pure strategies in role u_1 and three in role u_2. Here, the initial move by nature indicates $\rho(u_k, u_\ell) = \frac{1}{4}$ for all $1 \leq k, \ell \leq 2$. On the other hand, if $N = 1$, then $\rho(u_1, u_1) = 1$ and we have a symmetric game (e.g. only the left-hand subtree of Figure 9 formed by nature following the left-most direction at the root with probability 1). Similarly, if $N = 2$ and $\rho(u_1, u_2) = \rho(u_2, u_1) = \frac{1}{2}$, then $\rho(u_1, u_1) = \rho(u_2, u_2) = 0$ and so we have a bimatrix game (e.g. only the middle two subtrees of Figure 9 formed by nature following these two directions at the root with probability $\frac{1}{2}$). Thus, asymmetric games include both symmetric and bimatrix normal form games as special cases.

All asymmetric games have a single-stage extensive form representation with an initial move by nature and information sets $u_1, u_2, ..., u_N$ for both players. A pure strategy for player 1 specifies a choice at each of his information sets. It has the form e_i where $i = (i_1, ..., i_N)$ is a multi-index with i_k giving the choice of e_i at u_k. Each mixed strategy p is a discrete probability distribution over the finite set $\{e_i\}$ with weight p_i on e_i. This p induces a local behavior strategy p^k at each information set u_k given by

$$p_r^k = \sum_{\{i|i_k=r\}} p_i$$

and the Wright manifold is $W \equiv \{p \mid p_i = p_{(i_1,...,i_N)} = p_{i_1}^1 p_{i_2}^2 ... p_{i_N}^N\}$. W is invariant under the replicator equation.

2.1. Bimatrix games. Here, $N = 2$, $\rho(u_1, u_2) = \rho(u_2, u_1) = \frac{1}{2}$ and $\rho(u_1, u_1) = \rho(u_2, u_2) = 0$. By abuse of notation, let $A_{12} = A, A_{21} = B, p^1 = p, p^2 = q, e_i$ be the pure strategies in role u_1, and f_j be the pure strategies in role u_2. Then, on W,

$$\begin{aligned} \dot{p}_i &= p_i(e_i - p) \cdot Aq \\ \dot{q}_j &= q_j(f_j - q) \cdot Bp \end{aligned}$$

is the replicator equation restricted to W when the pure strategies are given as the appropriate unit vectors. This is the bimatrix replicator dynamics which we illustrate in Example 4.

Example 4. (Cressman, 2003) Consider the game with bimatrix

$$\begin{array}{c} \\ \mathrm{T} \\ \mathrm{I} \end{array} \begin{array}{c} \quad \mathrm{H} \qquad \mathrm{C} \\ \left[\begin{array}{cc} 5,4 & 1,6 \\ 4,0 & 3,-2 \end{array} \right] \end{array}.$$

There is no NE given by a pure strategy pair (e.g. at (T, H), player 2 does better by switching to C since $6 > 4$).[6] In fact, no strategy pair is a NE if either player uses a pure strategy. Thus, any NE (p^*, q^*) must be a completely mixed strategy for each player. In particular, there is a unique NE given by $(p_1^*, q_1^*) = \left(\frac{1}{2}, \frac{2}{3}\right)$ since

$$Aq^* = \left[\begin{array}{cc} 5 & 1 \\ 4 & 3 \end{array} \right] \left[\begin{array}{c} 2/3 \\ 1/3 \end{array} \right] = \left[\begin{array}{c} 11/3 \\ 11/3 \end{array} \right] \text{ and } Bp^* = \left[\begin{array}{cc} 4 & 0 \\ 6 & -2 \end{array} \right] \left[\begin{array}{c} 1/2 \\ 1/2 \end{array} \right] = \left[\begin{array}{c} 2 \\ 2 \end{array} \right].$$

However, this NE is not asymptotically stable since $H(p_1, q_1) \equiv p_1^2(1-p_1)^2 q_1^2(1-q_1)$ is a constant of motion under the replicator equation (i.e. $\frac{dH}{dt} = 0$) whose level curves are given in Figure 11. Trajectories of the replicator equation

$$\begin{aligned} \dot{p}_1 &= p_1(1 - p_1)(3q_1 - 2) \\ \dot{q}_1 &= q_1(1 - q_1)(2 - 4p_1) \end{aligned} \tag{2.1}$$

evolve clockwise around the interior rest point (p_1^*, q_1^*) along these curves.

Another way to see that $(p_1^*, q_1^*) = \left(\frac{1}{2}, \frac{2}{3}\right)$ is not asymptotically stable is to consider the time-adjusted replicator equation in the interior of the unit square

[6] This contrasts with Result 2 of Theorem 1 for perfect information games.

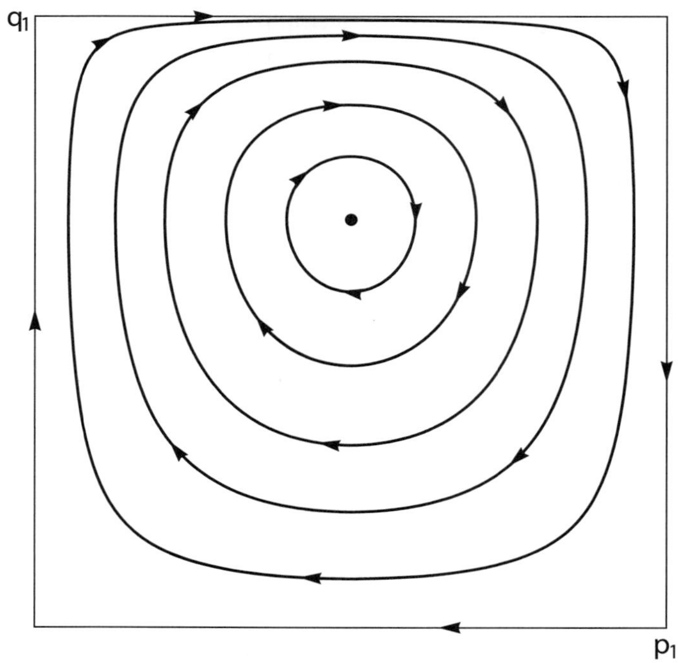

FIGURE 11. Trajectories of the bimatrix replicator equation for Example 4.

that divides the vector field in (2.1) by the Dulac function $p_1(1-p_1)q_1(1-q_1)$ to obtain

$$\begin{aligned} \dot{p}_1 &= \frac{(3q_1-2)}{q_1(1-q_1)} \\ \dot{q}_1 &= \frac{(2-4p_1)}{p_1(1-p_1)} \end{aligned} . \tag{2.2}$$

Trajectories of (2.2) are the same curves as those of (2.1) and evolve in the same direction. In particular, both dynamics have the same asymptotically stable interior points. Under the adjusted dynamics (2.2), a rectangle $\Delta p_1 \Delta q_1$ in the interior does not change area as it evolves since its horizontal and vertical cross-sections maintain the same lengths under this dynamics. (This invariance of area also follows from Liouville's result that "volumes" remain constant when the vector field is divergence free.) Thus no interior point can be asymptotically stable since no small rectangle containing it evolves to this point (i.e. to a region with zero area).

Example 4 is the well-known Buyer-Seller Game where a Buyer of some merchandise can either Trust the Seller to give an accurate value of this item or the Buyer can have the item inspected (i.e. play Inspect) to determine its true value. The Seller has a choice between Honest (give an accurate value) or Cheat (misrepresent its true value). The clockwise rotation of the trajectories in Figure 11 is not surprising given the cycling of the best responses to pure strategy pairs. What is not *a priori* clear is why trajectories cannot spiral in to the interior rest point making it asymptotically stable. This follows from the analysis above for the two-dimensional dynamics of Figure 11. It is also a consequence of part (a) of the following theorem for general bimatrix games (which can be proved using Liouville's result in higher dimensions).

Theorem 3. (Hofbauer and Sigmund, 1998) (a) A strategy pair (p^*, q^*) is an asymptotically stable rest point of the bimatrix replicator equation if and only if it is a strict NE. In particular, (p^*, q^*) is a pure strategy pair.
(b) If there is no interior NE, then all trajectories of the bimatrix replicator equation converge to the boundary.

2.2. Two-species ESS. Asymmetric games with two roles (i.e. $N = 2$) can be interpreted as games between two species by equating intraspecific interactions as between individuals playing a symmetric game in the same roles and interspecific interactions between individuals playing a bimatrix game in opposite roles. From this perspective, Figure 10 is then an example where there are both intra and inter specific interactions. On the other hand, bimatrix games such as the Buyer-Seller Game of Example 4 are then ones where all interactions are interspecific.

Suppose we extend Maynard Smith's original idea by saying that a (two-species) ESS is a monomorphic system with strategy pair (p^*, q^*) that cannot be successfully invaded by a rare (mutant) subsystem using a different strategy pair (p, q). That is, define (p^*, q^*) as a two-species ESS if it is asymptotically stable under the replicator equation based on the strategy pairs (p^*, q^*) and (p, q) whenever $(p, q) \neq (p^*, q^*)$. Suppose that A and D are the payoff matrices for intraspecific interactions (i.e. symmetric games) of species one and two respectively whereas B and C form the bimatrix game corresponding to interspecific interactions.

For the two-dimensional replicator equation based on the strategy pairs (p^*, q^*) and (p, q), let ε be the frequency of p in species one (so p^* has frequency $1 - \varepsilon$) and δ be the frequency of q in species two. The payoff of p is $p \cdot [A(\varepsilon p + (1-\varepsilon)\, p^*) + B(\delta q + (1-\delta)\, q^*)]$ and the average payoff in species one is $(\varepsilon p + (1-\varepsilon)\, p^*) \cdot [A(\varepsilon p + (1-\varepsilon)\, p^*) + B(\delta q + (1-\delta)\, q^*)]$. By the analogous expressions for the payoffs of species two, this replicator equation is
(2.3)

$$\begin{aligned} \dot{\varepsilon} &= (1-\varepsilon)\left((\varepsilon p + (1-\varepsilon)\, p^*) - p^*\right) \cdot [A(\varepsilon p + (1-\varepsilon)\, p^*) + B(\delta q + (1-\delta)\, q^*)] \\ \dot{\delta} &= (1-\delta)\left((\delta q + (1-\delta)\, q^*) - q^*\right) \cdot [C(\varepsilon p + (1-\varepsilon)\, p^*) + D(\delta q + (1-\delta)\, q^*)]. \end{aligned}$$

Then

Theorem 4. (Cressman, 2003) (a) (p^*, q^*) is a two-species ESS if and only if

$$\begin{array}{ll} \text{either} & p^* \cdot (Ap + Bq) > p \cdot (Ap + Bq) \\ \text{or} & q^* \cdot (Cp + Dq) > q \cdot (Cp + Dq) \end{array} \tag{2.4}$$

for all strategy pairs (p, q) that are sufficiently close (but not equal) to (p^*, q^*).
(b) If (p^*, q^*) is a two-species ESS, then it is asymptotically stable for the two-species replicator equation (i.e. based on all pure strategies of the asymmetric game).

PROOF. (a) Fix (p, q) with $p \neq p^*$ and $q \neq q^*$. Notice that the dynamics (2.3) leaves the unit square and each of its edges invariant. We claim that $(0, 0)$ is asymptotically stable if and only if

(2.5)

$$\begin{array}{ll} \text{either} & ((\varepsilon p + (1-\varepsilon)\, p^*) - p^*) \cdot [A(\varepsilon p + (1-\varepsilon)\, p^*) + B(\delta q + (1-\delta)\, q^*)] < 0 \\ \text{or} & ((\delta q + (1-\delta)\, q^*) - q^*) \cdot [C(\varepsilon p + (1-\varepsilon)\, p^*) + D(\delta q + (1-\delta)\, q^*)] < 0 \end{array}$$

for all nonnegative ε and δ with (ε, δ) sufficiently close (but not equal) to $(0,0)$. This result applied to all choices of (p,q) completes the proof of part (a).

If $(p-p^*) \cdot [Ap^* + Bq^*] \neq 0$ or $(q-q^*) \cdot [Cp^* + Dq^*] \neq 0$, it is straightforward to show $(0,0)$ is asymptotically stable if and only if $(p-p^*) \cdot [Ap^* + Bq^*] \leq 0$ and $(q-q^*) \cdot [Cp^* + Dq^*] \leq 0$ and that these last two inequalities hold if and only if (2.5) holds. Thus, for the remainder of the proof, assume that $(p-p^*) \cdot [Ap^* + Bq^*] = 0$ and $(q-q^*) \cdot [Cp^* + Dq^*] = 0$. Then (2.3) becomes

$$\begin{array}{rcl} \dot{\varepsilon} & = & \varepsilon\,(1-\varepsilon)\,(p-p^*) \cdot [A\varepsilon(p-p^*) + B\delta(q-q^*)] \\ \dot{\delta} & = & \delta\,(1-\delta)\,(q-q^*) \cdot [C\varepsilon(p-p^*) + D\delta(q-q^*)] \end{array}$$

and the two-species ESS condition (2.4) can be rewritten as

$$\begin{array}{ll} \text{either} & (p-p^*) \cdot [A(p-p^*) + B(q-q^*)] < 0 \\ \text{or} & (q-q^*) \cdot [C(p-p^*) + D(q-q^*)] < 0. \end{array}$$

Since the $\varepsilon-$axis (i.e. when $\delta = 0$) is invariant, $\varepsilon = 0$ is asymptotically stable on it if and only if $(p-p^*) \cdot A(p-p^*) < 0$. By the same argument applied to the $\delta-$axis, asymptotic stability of $(0,0)$ implies $(p-p^*) \cdot A(p-p^*) < 0$ and $(q-q^*) \cdot D(q-q^*) < 0$. Suppose that these two strict inequalities are true. If $(p-p^*) \cdot B(q-q^*) \leq 0$ or $(q-q^*) \cdot C(p-p^*) \leq 0$, then $(0,0)$ is asymptotically stable and (2.4) holds. On the other hand, if $(p-p^*) \cdot B(q-q^*) > 0$ and $(q-q^*) \cdot C(p-p^*) > 0$, then $\dot{\varepsilon} = 0$ on the line

$$\varepsilon = -\frac{(p-p^*) \cdot B(q-q^*)}{(p-p^*) \cdot A(p-p^*)}\delta$$

through the origin with positive slope and $\dot{\varepsilon} < 0$ below this $\varepsilon-$isocline. Similarly $\dot{\delta} = 0$ on the line

$$\varepsilon = -\frac{(q-q^*) \cdot D(q-q^*)}{(q-q^*) \cdot C(p-p^*)}\delta$$

through the origin with positive slope and $\dot{\delta} < 0$ above this $\delta-$isocline. From the phase diagram, it follows that $(0,0)$ is asymptotically stable if and only if the $\varepsilon-$isocline is steeper than the $\delta-$isocline. On the other hand, (2.4) holds if the $\varepsilon-$isocline is steeper than the $\delta-$isocline. That is, asymptotic stability of $(0,0)$ implies (2.4). Conversely, if (2.4) is not true for some choice of $(p,q) \neq (p^*, q^*)$, it is again clear from the phase diagram that $(0,0)$ is not asymptotically stable for the replicator equation based on these two strategy pairs.

(b) Suppose that species one (respectively, two) has n (respectively m) pure strategies. We only consider the case where (p^*, q^*) is in the interior of $\Delta^n \times \Delta^m$. The key to the stability of (p^*, q^*) under the replicator equation on $\Delta^n \times \Delta^m$ is that (2.4) is equivalent to the existence of an $r > 0$ such that

$$(p-p^*) \cdot [A(p-p^*) + B(q-q^*)] + r\,(q-q^*) \cdot [C(p-p^*) + D(q-q^*)] < 0$$

for all $(p,q) \in \Delta^n \times \Delta^m$ that are not equal to (p^*, q^*) (see Cressman, 2003). Consider the function

$$V(p,q) \equiv \Pi_{i=1}^{n} (p_i)^{p_i^*} \left(\Pi_{j=1}^{m} (q_j)^{q_j^*} \right)^r,$$

which is defined on $\Delta^n \times \Delta^m$ and has a global maximum at (p^*, q^*). It can then be shown that $\dot{V}(p,q) > 0$ if $(p,q) \neq (p^*, q^*)$ and (p,q) is in the interior of $\Delta^n \times \Delta^m$. That is, V is a strict Lyapunov function (Hofbauer and Sigmund, 1998) and so (p^*, q^*) is globally asymptotically stable. □

If there are no intraspecific interactions (take A and D as the zero matrices 0 of the appropriate size), then (p^*, q^*) is a two-species ESS if and only if it is a strict NE (e.g. by taking $q = q^*$, we have that $p^* \cdot Bq^* > p \cdot Bq^*$ if $p \neq p^*$ since $q^* \cdot Cp = q \cdot Cp$). That is, by Theorems 3 and 4, (p^*, q^*) is a two-species ESS for a bimatrix game if and only if it is a strict NE and this is true if and only if it is asymptotically stable under the bimatrix replicator equation.

At the other extreme, suppose that there are no interspecific interactions (take B and C as zero matrices). Then, (p^*, q^*) is a two-species ESS if and only p^* is a single-species ESS for species one and q^* is a single-species ESS for species two. For example, when $q = q^*$, we need $p^* \cdot Ap > p \cdot Ap$ for all p that are sufficiently close (but not equal) to p^*. Recall that this inequality condition, called local superiority by Weibull (1995), characterizes the single-species ESS (of species one). From this result, there can be two-species ESSs that are not strict NE (see Example 5 below). In particular, there can be completely mixed ESSs.

From these two extremes, we see that the concept of a two-species ESS combines and generalizes the concepts of single-species ESS of symmetric games and the strict NE of bimatrix games.

Example 5. (Krivan et al., 2008) Suppose that there are two species competing in two different habitats (or patches) and that the overall population size (i.e. density) of each species is fixed. Also assume that the fitness of an individual depends only on its species, the patch it is in and the density of both species in this patch. Then strategies of species one and two can be parameterized by the proportions p_1 and q_1 respectively of these species that are in patch one. If individual fitness (i.e. payoff) is positive when a patch is unoccupied and linearly decreasing in patch densities, it is of the form

$$\begin{aligned} F_i &= r_i\left(1 - \frac{p_i M}{K_i} - \frac{\alpha_i q_i N}{K_i}\right) \\ G_i &= s_i\left(1 - \frac{q_i N}{L_i} - \frac{\beta_i p_i M}{L_i}\right). \end{aligned}$$

Here, F_i is the fitness of a species one individual in patch i, G_i is the fitness of a species two individual in patch i, $p_2 = 1 - p_1$ and $q_2 = 1 - q_1$. All other parameters are fixed and positive (see Remark 7 below).

By linearity, these fitnesses can be represented by a two-species asymmetric game with payoff matrices

$$A = \begin{bmatrix} r_1 - \frac{r_1 M}{K_1} & r_1 \\ r_2 & r_2 - \frac{r_2 M}{K_2} \end{bmatrix} \qquad B = \begin{bmatrix} -\frac{\alpha_1 r_1 N}{K_1} & 0 \\ 0 & -\frac{\alpha_2 r_2 N}{K_2} \end{bmatrix}$$

$$C = \begin{bmatrix} -\frac{\beta_1 s_1 M}{L_1} & 0 \\ 0 & -\frac{\beta_2 s_2 M}{L_2} \end{bmatrix} \qquad D = \begin{bmatrix} s_1 - \frac{s_1 N}{L_1} & s_1 \\ s_2 & s_2 - \frac{s_2 N}{L_2} \end{bmatrix}.$$

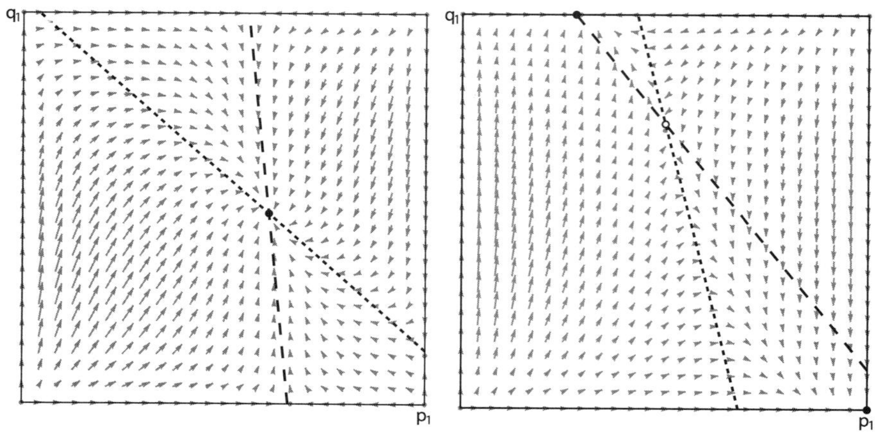

FIGURE 12. Vector fields for the two-patch Habitat Selection Game. The equal fitness lines of species one (dashed line) and species two (dotted line) intersect in the unit square. Solid dots are ESSs. (a) A unique ESS in the interior. (b) Two ESSs on the boundary.

For example, $F_i = e_i \cdot (Ap + Bq)$. At an equilibrium (p, q), all individuals present in species one must have the same fitness as do all individuals present in species two.

Suppose that both patches are occupied at the equilibrium (p, q). Then (p, q) is a NE and (p_1, q_1) is a point in the interior of the unit square that satisfies

$$\begin{aligned} r_1\left(1 - \frac{p_1 M}{K_1} - \frac{\alpha_1 q_1 N}{K_1}\right) &= r_2\left(1 - \frac{(1-p_1)M}{K_2} - \frac{\alpha_2(1-q_1)N}{K_2}\right) \\ s_1\left(1 - \frac{q_1 N}{L_1} - \frac{\beta_1 p_1 M}{L_1}\right) &= s_2\left(1 - \frac{(1-q_1)N}{L_2} - \frac{\beta_2(1-p_1)M}{L_2}\right). \end{aligned}$$

That is, these two "equal fitness" lines (which have negative slopes) intersect at (p_1, q_1) as in Figure 12.

The interior NE (p, q) is a two-species ESS if and only if the equal fitness line of species one is steeper than that of species two (cf. the proof of Theorem 4). That is, (p, q) is an interior two-species ESS in Figure 12A but not in Figure 12B. The interior two-species ESS in Figure 12A is globally asymptotically stable under the replicator equation.

Figure 12B has two two-species ESSs, both on the boundary of the unit square. One is a pure strategy pair strict NE with species one and two occupying separate patches $(p_1 = 1, q_1 = 0)$ and the other has species two in patch one and species one split between the two patches $(0 < p_1 < 1, q_1 = 1)$. Both are locally asymptotically stable under the replicator equation with basins of attraction formed by an invariant separatrix, joining the two vertices corresponding to both species in the same patch, on which trajectories evolve to the interior NE.

If the equal fitness lines do not intersect in the interior of the unit square, then there is exactly one two-species ESS. This is on the boundary (either a vertex or on an edge) and is globally asymptotically stable under the replicator equation.

Remark 7. Example 5 is called a (two-patch) Habitat Selection Game for two competitive species. The fixed parameters then have biological interpretations. Specifically, for species one, M is the total population size; r_i, K_i and α_i are its intrinsic growth rate, carrying capacity and interspecific competition coefficient (modelling the effect of the second species on the first) in patch i respectively. The analogous parameters for species two are N; s_i, L_i and β_i. Linearity of the fitness functions corresponds to Lotka-Volterra interactions.

Habitat selection games for a single species were already introduced before evolutionary game theory was developed when Fretwell and Lucas (1969) defined an ideal free distribution (IFD) to be a patch distribution whereby the fitness of any individual in an occupied patch was the same as the fitness of any other individual and at least as high as what would be the fitness in any unoccupied patch. If patch fitness is decreasing in patch density, then the IFD and ESS concepts are identical for a single species. In fact, there is a unique IFD and it is globally asymptotically stable under the replicator equation.

For two species, some authors consider an interior NE to be a (two-species) IFD. Example 5 shows such NE may be unstable (Figure 12B) and so justifies the perspective of others who restrict the IFD concept to two-species ESSs.

Remark 8. The generalization of Theorem 4 to three (or more) species is a difficult problem (Cressman et al., 2001). It is possible to characterize a monomorphic three-species ESS as one where, at all nearby strategy distributions, at least one species does better using its ESS strategy. However, such an ESS concept does not always imply stability of the three-species replicator equation that is based on the entire set of pure strategies for each species.

3. Games with Continuous Strategy Spaces

When players can choose from a continuum of pure strategies, the connections between NE and dynamic stability become more complicated. For example, the standard result forming one part of the Folk Theorem of Evolutionary Game Theory (Hofbauer and Sigmund, 1998) that a strict NE is asymptotically stable under the replicator equation (as well as most other deterministic evolutionary dynamics) is true for all games that have a finite set of pure strategies including the symmetric and asymmetric games of extensive or normal form in Sections 1 and 2. However, a strict NE is not always dynamically stable for games with continuous strategy spaces as seen in Example 6 below. Now, dynamic stability requires additional conditions such as that of a continuously stable strategy (CSS) or a neighborhood invader strategy (NIS) introduced by Eshel (1983) and Apaloo (1997) respectively. In fact, the exact requirement depends on the form of the evolutionary dynamics.

3.1. Symmetric games with a continuous strategy space. In this section, static conditions are developed for stability under the two standard dynamics for games with a continuous strategy space S; namely, adaptive dynamics and the replicator equation. The canonical equation of adaptive dynamics (Dieckmann and Law, 1996; Dercole and Rinaldi, 2008) models the evolution of the population mean strategy $x \in S$ by assuming that the population is always monomorphic at its mean. Then x evolves through trait substitution in a direction y of nearby mutants that can invade due to their higher payoff than x when playing against this monomorphism.

The replicator equation is now a dynamic on the space $\Delta(S)$ of Borel probability measures over the strategy space S (Bomze, 1991). This infinite-dimensional dynamical system restricts to the replicator equation of a symmetric normal form game when a finite subset of S is taken as the strategy set. From the perspective of the replicator equation that describes the evolution of the population strategy distribution $P \in \Delta(S)$ rather than the evolution of the population mean, the canonical equation becomes a heuristic tool that approximates how the mean evolves by ignoring effects due to the diversity of strategies in the population.

3.1.1. *One-dimensional strategy space.* Suppose that S is a convex compact subset of $\mathbf{R}$ (i.e. a closed and bounded interval). Following Hofbauer and Sigmund (1990) if the payoff to y interacting with x, $\pi(y,x)$, is continuously differentiable, then the canonical equation of adaptive dynamics has the form (up to a change in time scale)

$$\dot{x} = \frac{\partial \pi(y,x)}{\partial y}\Big|_{y=x} \tag{3.1}$$

at interior points of S (i.e. for $x \in \text{int}(S)$). That is, x increases if $\pi(y,x)$ is an increasing function of y for y close to x.

An $x^* \in \text{int}(S)$ is an equilibrium if $\pi_1(x^*,x^*) = 0$ (here π_1 is the partial derivative of π with respect to the first variable). Such an equilibrium is called convergence stable (Christiansen, 1991) if it is asymptotically stable under (3.1). If $\pi(y,x)$ has continuous partial derivatives up to second order, then x^* is convergence stable if

$$\frac{d}{dx}\left[\frac{\partial \pi(y,x)}{\partial y}\Big|_{y=x}\right]\Big|_{x=x^*} < 0. \tag{3.2}$$

If π_{ij} is the second order partial derivative of π with respect to i and j, then this inequality condition is $\pi_{11}(x^*,x^*) + \pi_{12}(x^*,x^*) \equiv \pi_{11} + \pi_{12} < 0$. Conversely, if x^* is convergence stable, then $\pi_{11} + \pi_{12} \leq 0$. The following example illustrates the NE and convergence stability concepts for quadratic payoff functions.

Example 6. Let $S = [-1,1]$ be the set of pure strategies and $\pi(x,y) = ax^2 + bxy$ be the payoff to x playing against y for all $x, y \in S$ where a and b are fixed real numbers. Then x^* in the interior of S is a NE (i.e. $\pi(x,x^*) \leq \pi(x^*,x^*)$ for all $x \in S$) if and only if $x^* = 0$ and $a \leq 0$. Also, $x^* = 0$ is a strict NE if and only if $a < 0$. These results exclude the degenerate case where $2a + b = 0$ and $\pi(x,y) = a(x-y)^2 - ay^2$. In this case (which we ignore from now on), every $x \in S$ is a strict NE if $a < 0$, a NE if $a = 0$, and there are no pure strategy NE if $a > 0$.

From (3.1), adaptive dynamics is now

$$\dot{x} = \frac{\partial \pi(y,x)}{\partial y}\Big|_{y=x} = (2a+b)x.$$

The only equilibrium is $x^* = 0$ and it is convergence stable if and only if $2a+b < 0$. In particular, $x^* = 0$ may be a strict NE but not convergence stable (e.g. $a < 0$ and $2a+b > 0$) or may be a convergence stable rest point that is not a NE (e.g. $a > 0$ and $2a+b < 0$). In the first case, the strict NE is a rest point of adaptive dynamics that is unattainable from nearby monomorphic populations. The population evolves to the endpoint of S closest to the initial value of x (e.g. x evolves to 1 if x is positive initially).

In the latter case, once $x^* = 0$ becomes established as the population monomorphism, it is vulnerable to invasion by mutants since $\pi(x, x^*) > \pi(x^*, x^*)$ for all nonzero $x \in S$. Alternatively, an $x \in S$ closer to x^* than y cannot invade a dimorphic population that is evenly split between the strategies $\pm y$ for $| y | > | x |$ since the expected payoff to x (namely, $\frac{1}{2}\pi(x, y) + \frac{1}{2}\pi(x, (-y))$) is less than the expected payoff to either y or $-y$. In fact, x can invade this dimorphism if and only if $| x | > | y |$. For either of these reasons, it is usually assumed that evolutionary stability (from the perspective of adaptive dynamics) of an $x^* \in S$ requires a convergence stable equilibrium that is also a strict NE (see the following Definition 1 and Theorem 5). A convergence stable rest point that is not a strict NE forms the basis of an initial evolutionary branching (Geritz et al., 1998; Doebeli and Dieckmann, 2000) into a dimorphic system.

There is some disagreement whether the strict NE condition for x^* should hold for all $x \in S$ or be restricted to those strategies close to x^*. In the following, we take the second approach and call these neighborhood strict NE to make this choice clear. Such NE are also called ESS (Marrow et al., 1996) or are said to satisfy the ESS Maximum Principle (Vincent and Brown, 2005). We will not use the ESS terminology in Section 3 since the meaning of ESS is not universally accepted for games with a continuous strategy space (Apaloo et al, 2009).

Definition 1. (Eshel, 1983) Suppose the strategy space S is a subinterval of real numbers. An $x^* \in S$ is a *neighborhood continuously stable strategy (CSS)* if there exists an $\varepsilon > 0$ such that, for all $x \in S$ with $0 < | x - x^* | < \varepsilon$, the following two conditions hold.

(i) $\pi(x, x^*) < \pi(x^*, x^*)$ (neighborhood strict NE condition).

(ii) There exists $\eta > 0$ (which depends on x) such that, for all $x' \in S$ with $0 < | x' - x | < \eta$, $\pi(x', x) > \pi(x, x)$ if and only if $| x' - x^* | < | x - x^* |$ (convergence stability condition).

The convergence stability condition of Definition 1 (ii) is used in Theorem 5 (b). The proof there shows it has the same relationship to $\pi_{11} + \pi_{12}$ as the dynamic approach to convergence stability through inequality (3.2).

The CSS concept,like the ESS definition of Maynard Smith (1982), is defined in terms of static payoff comparisons that are meant to predict the outcome of evolutionary dynamics. For the CSS (as well as the related notion of neighborhood half superiority given in Definition 2), the connection to dynamic stability under the canonical equation is summarized in Theorem 5. Neighborhood superiority for a pure strategy requires the extension of the payoff functions to distributions $P \in \Delta(S)$. Let δ_x be the Dirac delta distribution with full weight on the point $x \in S$ and assume that individuals interact in random pairwise contests. Then $\pi(\delta_x, P) \equiv \int_S \pi(x, y)P(dy)$ (also denoted by $\pi(x, P)$) is the expected payoff to an individual who uses strategy x if the population strategy distribution is P and $\pi(P, P) \equiv \int_S \pi(x, P)P(dx)$ is the mean payoff of the population (Bomze and Pötscher, 1989).

The following definition is given for general multi-dimensional strategy spaces. The support of P is the closed set given by $\{x \in S \mid P(\{y : | y - x | > \varepsilon\}) > 0$ for all $\varepsilon > 0\}$.

Definition 2. (Cressman, 2009) Suppose the strategy space S of a symmetric game is a subset of $\mathbf{R}^n$ and $0 \leq p^* < 1$ is fixed. Strategy $x^* \in S$ is *neighborhood*

p^*-superior if $\pi(x^*, P) > \pi(P, P)$ for all $P \in \Delta(S)$ with $1 > P(\{x^*\}) \geq p^*$ and the support of P sufficiently close to x^*. It is *neighborhood superior* (respectively, *neighborhood half-superior*) if $p^* = 0$ (respectively, $p^* = \frac{1}{2}$). Strategy $x^* \in S$ is *globally p^*-superior* if $\pi(x^*, P) > \pi(P, P)$ for all $P \in \Delta(S)$ with $1 > P(\{x^*\}) \geq p^*$

Theorem 5. Suppose that S is one dimensional and $x^* \in$int(S) is a rest point of adaptive dynamics (3.1) (i.e. $\pi_1(x^*, x^*) = 0$).
(a) If $\pi_{11} < 0$, then x^* is a neighborhood strict NE. Conversely, if x^* is a neighborhood strict NE, then $\pi_{11} \leq 0$.
(b) If $\pi_{11} + \pi_{12} < 0$, then x^* is convergence stable. Conversely, if x^* is convergence stable, then $\pi_{11} + \pi_{12} \leq 0$.
(c) If $\pi_{11} < 0$ and $\pi_{11} + \pi_{12} < 0$, then x^* is a neighborhood CSS and neighborhood half-superior. Conversely, if x^* is a neighborhood CSS or neighborhood half-superior, then $\pi_{11} \leq 0$ and $\pi_{11} + \pi_{12} \leq 0$.

PROOF. These results follow from the Taylor expansion of $\pi(x, y)$ about (x^*, x^*); namely,

$$\begin{aligned}\pi(x,y) &= \pi(x^*,x^*) + \pi_1(x^*,x^*)(x-x^*) + \pi_2(x^*,x^*)(y-x^*)\\ &\quad + \frac{1}{2}\left[\pi_{11}(x-x^*)^2 + 2\pi_{12}(x-x^*)(y-x^*) + \pi_{22}(y-x^*)^2\right]\\ &\quad + \text{ higher order terms.}\end{aligned}$$

For $x^* \in$int(S), $\dot{x} = \pi_1(x^*, x^*)$ at $x = x^*$ and so $\pi_1 = 0$ since x^* is a rest point of (3.1).

(a) From the Taylor expansion, $\pi(x, x^*) - \pi(x^*, x^*) = \frac{1}{2}\pi_{11}(x - x^*)^2$ up to second order terms. Thus, x^* is a neighborhood strict NE (i.e. $\pi(x, x^*) < \pi(x^*, x^*)$ when x is sufficiently close (but not equal) to x^*) if $\pi_{11} < 0$. Conversely, if $\pi_{11} > 0$, x^* is not a neighborhood strict NE.

(b) From the Taylor expansion, $\pi(x', x) - \pi(x, x)$ is given by

$$\begin{aligned}&\pi_2(x^*,x^*)(x-x^*) + \frac{1}{2}\left[\pi_{11}(x'-x^*)^2 + 2\pi_{12}(x'-x^*)(x-x^*) + \pi_{22}(x-x^*)^2\right]\\ &-\left[\pi_2(x^*,x^*)(x-x^*) + \frac{1}{2}\left[\pi_{11}(x-x^*)^2 + 2\pi_{12}(x-x^*)^2 + \pi_{22}(x-x^*)^2\right]\right]\\ =\;& \frac{1}{2}\pi_{11}\left((x'-x^*)^2 - (x-x^*)^2\right) + \pi_{12}(x'-x^* - (x-x^*))(x-x^*)\\ =\;& (x'-x)\left[\left(\frac{x'+x}{2} - x^*\right)\pi_{11} + (x-x^*)\pi_{12}\right]\end{aligned}$$

up to second order terms. If $| x' - x | < \eta$ with $\eta <| x - x^* |$ small, then $\frac{x'+x}{2} \cong x$ and so $\pi(x', x) - \pi(x, x) \cong (x' - x)(x - x^*)(\pi_{11} + \pi_{12})$. Suppose $| x' - x^* | <| x - x^* |$ (i.e. x' is closer to x^* than x). Then $(x' - x)(x - x^*) < 0$ and so $\pi(x', x) - \pi(x, x) > 0$ if $\pi_{11} + \pi_{12} < 0$ and $\pi(x', x) - \pi(x, x) < 0$ if $\pi_{11} + \pi_{12} > 0$.

(c) Parts (a) and (b) combine to show that x^* is a neighborhood CSS if $\pi_{11} < 0$ and $\pi_{11} + \pi_{12} < 0$ and that, if x^* is a neighborhood CSS, then $\pi_{11} \leq 0$ and $\pi_{11} + \pi_{12} \leq 0$. Assume that x^* is neighborhood half-superior. With $P = \frac{1}{2}\delta_x + \frac{1}{2}\delta_{x^*}$,

$P(\{x^*\}) = \frac{1}{2}$ and

$$\begin{aligned}\pi(x^*,P)-\pi(P,P) &= \frac{1}{2}\left[\pi(x^*,x)+\pi(x^*,x^*)\right] \\ &\quad -\frac{1}{4}\left[\pi(x,x)+\pi(x,x^*)+\pi(x^*,x)+\pi(x^*,x^*)\right] \\ &= \frac{1}{4}\left[\pi(x^*,x^*)+\pi(x^*,x)-\pi(x,x^*)-\pi(x,x)\right] \\ &\cong -\frac{1}{4}\left(\pi_{11}+\pi_{12}\right)(x-x^*)^2.\end{aligned}$$

Thus, $\pi_{11}+\pi_{12} \leq 0$ since $\pi(x^*,P)-\pi(P,P) > 0$. Now take $P = \varepsilon\delta_x + (1-\varepsilon)\delta_{x^*}$. A similar calculation yields $\pi(x^*,P)-\pi(P,P) \cong \varepsilon\left(\pi(x^*,x^*)-\pi(x,x^*)\right)$ up to linear terms in ε. Thus, x^* is a neighborhood strict NE. For the converse statements involving neighborhood half-superiority, see Cressman (2009). □

Except in borderline cases when $\pi_{11} = 0$ or $\pi_{11}+\pi_{12} = 0$, Theorem 5 characterizes both the stability of interior neighborhood strict NE under adaptive dynamics and those $x^* \in \mathrm{int}(S)$ that are neighborhood half-superior (i.e. $p^* = \frac{1}{2}$). On the other hand, a neighborhood CSS need not be stable under the replicator equation. This is shown in the continuation of Example 6 below that follows a brief development of the essential properties of this latter dynamics.

A trajectory P_t for $t \geq 0$ is a solution of the replicator equation if the weight $P_t(B)$ assigned to any Borel subset B of S satisfies

$$\dot{P}_t(B) = \int_B \left(\pi(\delta_x, P_t) - \pi(P_t,P_t)\right) P_t(dx). \tag{3.3}$$

If π is continuous, there is a unique solution given any initial $P_0 \in \Delta(S)$ (Oechssler and Riedel, 2001). Stability under this replicator equation is typically analyzed with respect to the weak topology for $\Delta(S)$. We are most interested in the dynamic stability of a monomorphic population (i.e. of δ_{x^*} for some $x^* \in S$) when x^* is in the support of P_0. Every neighborhood of δ_{x^*} in the weak topology contains the set of all distributions P whose support is within ε of x^* (i.e. $P(\{x : \mid x - x^* \mid > \varepsilon\}) = 0$) for some $\varepsilon > 0$ and so δ_x is in this neighborhood if $\mid x - x^* \mid < \varepsilon$.

Example 6. (Continued) Consider Example 6 again where now all individuals play either $x^* = 0$ or a nearby strategy x (which is fixed). For this restricted two-strategy game, the replicator equation becomes the one-dimensional dynamics $\dot{p} = p(1-p)(a+bp)x^2$ (here p is the frequency of strategy x) corresponding to the symmetric game whose normal form is

$$\begin{array}{c} \\ x \\ x^* \end{array} \begin{array}{c} \begin{array}{cc} x & \quad x^* \end{array} \\ \left[\begin{array}{cc} (a+b)x^2 & ax^2 \\ 0 & 0 \end{array}\right] \end{array}.$$

Take $a = -2$ and $b = 3$ so that $x^* = 0$ is a CSS and also half-superior (i.e. $a < 0$ and $2a + b < 0$). Since $a + b > 0$ and $a < 0$, both pure strategies are strict NE and so locally asymptotically stable for the replicator equation applied to this two-strategy game. However, neither is asymptotically stable for the infinite-dimensional replicator equation for the full game since, in the weak topology on $\Delta(S)$, any neighborhood of x^* includes all probability measures whose support is sufficiently close to x^*. Thus, asymptotic stability of x^* requires that x^* is globally stable for the two-strategy game and so $a+b \leq 0$. Instability in this example results from $a+b=1>0$.

In the following section, we extend the CSS concept to multi-dimensional strategy spaces and examine stability conditions with respect to the replicator equation.

3.1.2. *Multi-dimensional strategy space.* The above one-dimensional model and theory can be extended to multi-dimensional strategy spaces S that are compact convex subsets of $\mathbf{R}^n$ with $x^* \in S$ in its interior. Following the static approach of Lessard (1990), x^* is a neighborhood CSS if it is a neighborhood strict NE that satisfies condition (ii) of Definition 1 along each line through x^*. Theorem 6 then generalizes Theorem 5 in terms of the Taylor expansion about (x^*, x^*) of the payoff function

$$\begin{aligned} \pi(x,y) &= \pi(x^*,x^*) + \nabla_1\pi(x^*,x^*)(x-x^*) + \nabla_2\pi(x^*,x^*)(y-x^*) \\ &\quad + \frac{1}{2}\left[(x-x^*)\cdot A(x-x^*) + 2(x-x^*)\cdot B(y-x^*) + (y-x^*)\cdot C(y-x^*)\right] \\ &\quad + \text{higher order terms.} \end{aligned}$$

Here, ∇_1 and ∇_2 are gradient vectors with respect to x and y respectively (e.g. the ith component of $\nabla_1\pi(x^*,x^*)$ is $\frac{\partial\pi(x',x^*)}{\partial x'_i}\mid_{x'=x^*}$) and A, B, C are the $n \times n$ matrices with ij^{th} entries (all partial derivatives are evaluated at x^*)

$$A_{ij} \equiv \left[\frac{\partial^2}{\partial x'_j \partial x'_i}\pi(x',x^*)\right]; B_{ij} \equiv \left[\frac{\partial}{\partial x'_i}\frac{\partial}{\partial x_j}\pi(x',x)\right]; C_{ij} \equiv \left[\frac{\partial}{\partial x'_j}\frac{\partial}{\partial x'_i}\pi(x^*,x')\right].$$

An $n \times n$ matrix M is negative definite (respectively, negative semi-definite) if, for all nonzero $x \in \mathbf{R}^n$, $x \cdot Mx < 0$ (respectively, $x \cdot Mx \leq 0$).

Adaptive dynamics for multi-dimensional strategy spaces generalizing (3.1) now has the form

$$\dot{x} = C_1(x)\nabla_1\pi(y,x)\mid_{y=x} \tag{3.4}$$

where $C_1(x)$ is an $n \times n$ covariance matrix modeling the mutation process (and its rate) in different directions (Leimar, 2009). We will assume that $C_1(x)$ is a positive-definite symmetric matrix for $x \in$int(S) that depends continuously on x. System (3.4) is called the canonical equation of adaptive dynamics (when S is multi-dimensional).

Theorem 6. (Cressman, 2009; Leimar, 2009) Suppose $x^* \in$int(S) is a rest point of (3.4) (i.e. $\nabla_1\pi(x^*,x^*)=0$).
(a) If A is negative definite, then x^* is a neighborhood strict NE. Conversely, if x^* is a neighborhood strict NE, then A is negative semi-definite.
(b) If $A+B$ is negative definite, then x^* is convergence stable for any choice of covariance matrix $C_1(x)$ (i.e. x^* is an asymptotically stable rest point of (3.4)). Conversely, if x^* is convergence stable for any choice of covariance matrix $C_1(x)$, then $A+B$ is negative semi-definite.
(c) If A and $A+B$ are negative definite, then x^* is a neighborhood CSS and neighborhood half-superior. Conversely, if x^* is a neighborhood CSS or neighborhood half-superior, then A and $A+B$ are negative semi-definite.

PROOF. The proofs of parts (a) and (c) are similar to the corresponding proofs of Theorem 5.

(b) Suppose $A+B$ is negative definite and $C_1(x)$ is a positive definite symmetric matrix that depends continuously on x. From the Taylor expansion of $\pi(x,y)$, the canonical equation (3.4) has the form

$$\dot{x} = C_1\,(A+B)\,(x-x^*) + \text{ higher order terms}$$

where $C_1 = C_1(x^*)$. Let $V(x) \equiv \left(C_1^{-1}(x-x^*)\right)\cdot(x-x^*)$. Since C_1^{-1} is also positive definite and symmetric, $V(x) \geq 0$ for all $x \in \mathbf{R}^n$ with equality only at $x = x^*$. Furthermore

$$\begin{aligned} \dot{V}(x) &= (C_1^{-1}\dot{x})\cdot(x-x^*) + \left(C_1^{-1}(x-x^*)\right)\cdot\dot{x} + \text{ higher order terms} \\ &\cong 2C_1^{-1}C_1\,(A+B)\,(x-x^*)\cdot(x-x^*) \\ &= 2(x-x^*)\cdot(A+B)\,(x-x^*). \end{aligned}$$

Thus, for x sufficiently close (but not equal) to x, $\dot{V}(x) < 0$. That is, V is a local Lyapunov function (Hofbauer and Sigmund, 1998) and so x^* is asymptotically stable.

Conversely, suppose that x^* is convergence stable. If $A+B$ is not negative semi-definite, then there exists an $x \in \mathbf{R}^n$ such that $x\cdot(A+B)\,x > 0$. Let $\widehat{C}_1$ be the orthogonal projection of $\mathbf{R}^n$ onto the line $\{ax \mid a \in \mathbf{R}\}$ through the origin and x. $\widehat{C}_1$ is a positive semi-definite symmetric matrix. The line $\{x^* + ax \mid a \in \mathbf{R}\}$ is invariant under the canonical equation when $C_1(x) = \widehat{C}_1$ and the linearization at x^* is

$$\dot{a} = (x\cdot(A+B)\,x)\,a.$$

Thus the canonical equation with respect to $\widehat{C}_1$ has the positive eigenvalue $x\cdot(A+B)\,x$ at x^*. For any positive definite covariance matrix C_1 sufficiently close to $\widehat{C}_1$, the linearization still has an eigenvalue with positive real part. Since x^* is not stable for this choice of C_1, x^* is not convergence stable (a contradiction). □

In the continuation of Example 6, we saw that dynamic stability with respect to the replicator equation requires more than the CSS concept. In Theorem 7, it is NIS that takes the place of the convergence stability CSS condition and attractivity replaces dynamics stability for the replicator equation. These are defined as follows.

Definition 3. (a) (Apaloo, 1997)[7] Suppose the strategy space S of a symmetric game is a subset of $\mathbf{R}^n$. Strategy $x^* \in S$ is a *neighborhood invader strategy (NIS)* if $\pi(x^*,x) > \pi(x,x)$ for all $x \in S$ sufficiently close (but not equal) to x^*.
(b) (Cressman et al., 2006) δ_{x^*} is *neighborhood attracting* under the replicator equation if, for any initial distribution P_0 with support sufficiently close to x^* and $P_0(\{x^*\}) > 0$, P_t converges to δ_{x^*} in the weak topology.

Theorem 7. (Cressman et al., 2006) Suppose x^* is in the interior of S and satisfies $\nabla_1\pi(x^*,x^*) = 0$.
(a) If $A+2B$ is negative definite, then x^* is a NIS. Conversely, if x^* is a NIS, then $A+2B$ is negative semi-definite.

---

[7] A NIS is also called a "good invader" strategy (Kisdi and Meszéna, 1995) and an "invading when rare" strategy (Courteau and Lessard, 2000).

(b) If A and $A+2B$ are negative definite, then x^* is neighborhood superior and neighborhood attracting under the replicator equation. Conversely, if x^* is neighborhood superior or neighborhood attracting, then A and $A+2B$ are negative semi-definite.

PROOF. (a) From the Taylor expansion,

$$\begin{aligned}\pi(x^*,x)-\pi(x,x) &= \nabla_1\pi(x-x^*)+\frac{1}{2}[x-x^*)\cdot C(x-x^*)] \\ &\quad -\left[\nabla_1\pi(x-x^*)+\frac{1}{2}[(x-x^*)\cdot(A+2B+C)(x-x^*)]\right] \\ &\quad + \text{higher order terms.} \\ &\cong -\frac{1}{2}(x-x^*)\cdot(A+2B)(x-x^*).\end{aligned}$$

Thus, x^* is a NIS if $A+2B$ is negative definite. Conversely, if $A+2B$ is not negative semi-definite, then there is an $x\in S$ arbitrarily close to x^* such that $\pi(x^*,x)-\pi(x,x)<0$ (i.e. x^* is not a NIS).

(b) The proof of the relationship between neighborhood superiority and the matrices A and $A+2B$ follows the same steps as the proof of Theorem 5 (c). Specifically, the negative definiteness of A is considered through strategy distributions of the form $P=\varepsilon\delta_x+(1-\varepsilon)\delta_{x^*}$. On the other hand, for negative definiteness of $A+2B$, consider $P=\delta_x$ for x close to x^*. Then $\pi(x^*,P)-\pi(P,P)=\pi(x^*,x)-\pi(x,x)$ and this is related to $A+2B$ by part (a).

Now suppose that A and $A+2B$ are negative definite. From (3.3),

$$\dot{P}_t(\{x^*\})=(\pi(\delta_{x^*},P_t)-\pi(P_t,P_t))\,P_t(\{x^*\}).$$

Since the support of P_t is invariant under the replicator equation (Oechssler and Riedel, 2001), $P_t(\{x^*\})$ is increasing if P_0 has support sufficiently close to x^*. From this it follows that P_t converges to δ_{x^*} in the weak topology (see Cressman (2010) for details of this and of the converse statement). □

Remark 9. The analysis in Cressman et al. (2006) shows that one must be careful in extending the statements of Theorem 7, part b, from neighborhood attractivity to asymptotic stability or from $P_0(\{x^*\})>0$ to x^* in the support of P_0, especially if the payoff function is not symmetric (i.e. $\pi(x,y)\neq\pi(y,x)$ for some $x,y\in S$). In fact, there remain open problems in these cases. On the other hand, there are examples that show neither negative definiteness nor negative semi-definiteness provide complete characterizations in any part of Theorems 6 and 7. For example, there are borderline cases with x^* a neighborhood strict NE and $A+2B$ negative semi-definite for which x^* is a NIS in one case but not in the other (Cressman et al., 2006).

3.2. Asymmetric games with continuous strategy spaces. The above theory of multi-dimensional CSS and NIS as well as their connections to evolutionary dynamics have been extended to asymmetric games with continuous strategy spaces (Cressman, 2009, 2010). When there are two roles, it is shown there that the CSS and NIS can be characterized (excluding borderline cases) by payoff comparisons similar to those found for the two-species ESS when both roles have a finite number of strategies (see Theorem 4 and Definition 4). In this section, we will assume that the continuous strategy sets S and T for the two roles are both one-dimensional compact intervals and that payoff functions have continuous partial derivatives up to second order in order to avoid technical and/or notational complications.

For $(x,y) \in S \times T$, let $\pi_1(x';x,y)$ (respectively, $\pi_2(y';x,y)$) be the payoff to a player in role 1 (respectively, in role 2) using strategy $x' \in S$ (respectively $y' \in T$) when the population is monomorphic at (x,y). Note that π_1 has a different meaning here than in Section 3.1 where it was used to denote a partial derivative. With this terminology, the canonical equation of adaptive dynamics (c.f. (3.1)) becomes

$$\begin{aligned} \dot{x} &= k_1(x,y)\frac{\partial}{\partial x'}\pi_1(x';x,y)\mid_{x'=x} \\ \dot{y} &= k_2(x,y)\frac{\partial}{\partial y'}\pi_2(y';x,y)\mid_{y'=y} \end{aligned} \tag{3.5}$$

where $k_i(x,y)$ for $i=1,2$ are positive continuous functions of (x,y). At an interior rest point (x^*,y^*) of (3.5),

$$\frac{\partial \pi_1}{\partial x'} = \frac{\partial \pi_2}{\partial y'} = 0.$$

In particular, if (x^*,y^*) is a neighborhood strict NE (i.e. if $\pi_1(x;x^*,y^*) < \pi_1(x^*;x^*,y^*)$ and $\pi_2(y;x^*,y^*) < \pi_2(y^*;x^*,y^*)$ for all x and y sufficiently close but not equal to x and y respectively) in the interior of $S \times T$, then it is a rest point (x^*,y^*) of (3.5). (x^*,y^*) is called convergence stable (or strongly convergence stable as in Leimar, 2009) if it is asymptotically stable under (3.5) for any choice of k_1 and k_2.

The characterizations of these concepts in the following theorem are given in terms of the linearization of (3.5) about (x^*,y^*); namely,

$$\begin{bmatrix} \dot{x} \\ \dot{y} \end{bmatrix} = \begin{bmatrix} k_1(x^*,y^*) & 0 \\ 0 & k_2(x^*,y^*) \end{bmatrix} \begin{bmatrix} A+B & C \\ D & E+F \end{bmatrix} \begin{bmatrix} x-x^* \\ y-y^* \end{bmatrix} \tag{3.6}$$

where

$$\begin{aligned} A &\equiv \frac{\partial^2}{\partial x' \partial x'}\pi_1(x';x^*,y^*); B \equiv \frac{\partial}{\partial x'}\frac{\partial}{\partial x}\pi_1(x';x,y^*); C \equiv \frac{\partial}{\partial x'}\frac{\partial}{\partial y}\pi_1(x';x^*,y) \\ D &\equiv \frac{\partial}{\partial y'}\frac{\partial}{\partial x}\pi_2(y';x,y^*); E \equiv \frac{\partial}{\partial y'}\frac{\partial}{\partial y}\pi_2(y';x^*,y); F \equiv \frac{\partial^2}{\partial y' \partial y'}\pi_2(y';x^*,y^*) \end{aligned}$$

and all partial derivatives are evaluated at the equilibrium.

Theorem 8. (Cressman, 2010) Suppose (x^*,y^*) is a rest point (x^*,y^*) of (3.5) in the interior of $S \times T$.
(a) (x^*,y^*) is a neighborhood strict NE if A and F are negative. Conversely, if (x^*,y^*) is a neighborhood NE, then A and F are non-positive.
(b) (x^*,y^*) is convergence stable if, for all nonzero $(x,y) \in \mathbf{R}^2$, either $x((A+B)x + Cy) < 0$ or $y(Dx+(E+F)y) < 0$. Conversely, if (x^*,y^*) is convergence stable, then either $x((A+B)x+Cy) \leq 0$ or $y(Dx+(E+F)y) \leq 0$ for all $(x,y) \in \mathbf{R}^2$.
(c) (x^*,y^*) is convergence stable if $A+B<0, E+F<0$ and $(A+B)(E+F) > CD$. Conversely, if (x^*,y^*) is convergence stable, then $A+B \leq 0, E+F \leq 0$ and $(A+B)(E+F) \geq CD$.

PROOF. (a) These statements are straightforward consequences of the Taylor expansion of the payoff functions $\pi_1(x';x,y)$ and $\pi_2(y';x,y)$ about (x^*,y^*).

(b) (x^*, y^*) is convergence stable if both eigenvalues of the linearization (3.6) have negative real parts for any choice of positive $k_1(x^*, y^*)$ and $k_2(x^*, y^*)$. This latter condition holds if and only if the trace is negative (i.e. $k_1(x^*, y^*)(A+B) + k_2(x^*, y^*)(E+F) < 0$) and the determinant is positive (i.e. $k_1(x^*, y^*)k_2(x^*, y^*)[(A+B)(E+F) - DC] > 0$).

Assume that either $x((A+B)x + Cy) < 0$ or $y(Dx + (E+F)y) < 0$ for all nonzero $(x, y) \in \mathbf{R}^2$. In particular, with $(x, y) = (x, 0)$, we have $A + B < 0$. Analogously $E + F < 0$ and so the trace is negative. For a fixed nonzero y, let $x \equiv -\frac{C}{A+B}y$. Then $(A+B)x + Cy = 0$ and so $y(Dx + (E+F)y) < 0$. That is,

$$y\left(-\frac{CD}{A+B}y + (E+F)y\right) = \frac{(A+B)(E+F) - CD}{A+B}y^2$$

is negative and this implies the determinant is positive. Thus, (x^*, y^*) is convergence stable.

Conversely, assume that (x^*, y^*) is convergence stable. Then, the trace must be non-positive and the determinant non-negative for any choice of positive $k_1(x^*, y^*)$ and $k_2(x^*, y^*)$ (otherwise, there is an eigenvalue with positive real part). In particular, $A + B \leq 0$ and $E + F \leq 0$.

Case 1. If $CD \leq 0$, then either $xCy \leq 0$ or $yDx \leq 0$. Thus, either $x((A+B)x + Cy) \leq 0$ or $y(Dx + (E+F)y) \leq 0$ for all $(x, y) \in \mathbf{R}^2$.

Case 2. If $CD > 0$, we may assume without loss of generality that $C > 0$ and $D > 0$. Suppose that $x((A+B)x + Cy) > 0$. Then $xy > -\frac{(A+B)x^2}{C} > 0$. Thus

$$\begin{aligned} y(Dx + (E+F)y) &= \frac{y}{x}(Dx^2 + (E+F)xy) \\ &< \frac{y}{x}(Dx^2 - \frac{(A+B)(E+F)x^2}{C}) \leq 0. \end{aligned}$$

(c) These statements follow from the arguments used to prove part b. □

As in Section 3.1 for symmetric games, a neighborhood CSS is a neighborhood strict NE that is convergence stable (when borderline cases are excluded). For one-dimensional strategy spaces, S and T, parts b and c of Theorem 8 give equivalent conditions for convergence stability. Although the inequalities in part c are the easiest to use in practical examples, it is the approach in part b that is most directly tied to the theory of CSS, NIS and neighborhood superiority as well as their connections to evolutionary dynamics, especially as the strategy spaces become multi-dimensional. It is again neighborhood superiority according to part a of the following definition that unifies this theory (see Theorem 9 that assumes borderline cases are excluded).

Definition 4. (Cressman, 2010) Suppose (x^*, y^*) is in the interior of $S \times T$.
(a) Fix $0 \leq p^* < 1$. Strategy pair (x^*, y^*) is *neighborhood* p^*-*superior* if

$$\text{(3.7)} \qquad \text{either } \pi_1(x^*; P, Q) > \pi_1(P; P, Q) \text{ or } \pi_2(y^*; P, Q) > \pi_2(Q; P, Q)$$

for all $(P, Q) \in \Delta(S) \times \Delta(T)$ with $1 \geq P(\{x^*\}) \geq p^*$, $1 \geq Q(\{y^*\}) \geq p^*$ and the support of (P, Q) sufficiently close (but not equal) to (x^*, y^*). (x^*, y^*) is *neighborhood half-superior* if $p^* = \frac{1}{2}$. (x^*, y^*) is *neighborhood superior* if $p^* = 0$. (x^*, y^*) is *(globally)* p^*-*superior* if the support of (P, Q) in (3.7) is an arbitrary subset of $S \times T$ (other than $\{(x^*, y^*)\}$).

(b) Strategy pair (x^*, y^*) is a *neighborhood invader strategy (NIS)* if, for all (x, y) sufficiently close (but not equal) to (x^*, y^*), either $\pi_1(x^*; x, y) > \pi_1(x; x, y)$ or $\pi_2(y^*; x, y) > \pi_2(y; x, y)$.

Theorem 9. (Cressman, 2010) Suppose that (x^*, y^*) is in the interior of $S \times T$.
(a) (x^*, y^*) is a neighborhood CSS if and only if it is neighborhood half-superior.
(b) (x^*, y^*) is a neighborhood strict NE and NIS if and only if it is neighborhood superior.
(c) Consider evolution under the replicator equation generalizing (3.3) to asymmetric games and initial population distributions $(P_0, Q_0) \in \Delta(S) \times \Delta(T)$ that satisfy $P_0(\{x^*\})Q_0(\{y^*\}) > 0$. If (x^*, y^*) is a strict neighborhood NE and a NIS, then (P_t, Q_t) converges to $(\delta_{x^*}, \delta_{y^*})$ in the weak topology whenever the support of (P_0, Q_0) is sufficiently close to (x^*, y^*). Conversely, if (P_t, Q_t) converges to $(\delta_{x^*}, \delta_{y^*})$ in the weak topology for every (P_0, Q_0) with support sufficiently close to (x^*, y^*), then (x^*, y^*) is a neighborhood strict NE and NIS.

Theorem 8 is the (two-role) asymmetric counterpart of Theorem 5 for symmetric games when the continuous strategy spaces are one dimensional. Definition 4 and Theorem 9 generalize Definition 3 and Theorems 6 and 7 of Section 3.1 to these asymmetric games. Based on a thorough analysis of the Taylor expansions of the two payoff functions, their statements remain correct when S and T are multi-dimensional.

4. Conclusion

The static payoff comparisons (e.g. the ESS conditions) introduced by Maynard Smith (1982) to predict the behavioral outcome of evolution in symmetric games with finitely many strategies have been extended in many directions during the intervening years. These include biological extensions to multiple species and to population games as well as the equally important extensions to predict rational individual behavior in human conflict situations. As is apparent from this chapter, there is a complex relationship between these static conditions and evolutionary stability of the underlying dynamical system.

This chapter has emphasized evolutionary stability in (symmetric or asymmetric) extensive form games and games with continuous strategy spaces under the deterministic replicator equation that is based on random pairwise interactions. Evolutionary stability is also of much current interest for other game-theoretic models such as those that incorporate stochastic effects due to finite populations; models with assortative (i.e. non-random) interactions (e.g. games on graphs); models with multi-player interactions (e.g. public goods games). As the evolutionary theory behind these (and other) models is a rapidly expanding area of current research, it is impossible to know in what guise the evolutionary stability conditions will emerge in future applications. On the other hand, it is certain that Maynard Smith's original idea will continue to play a central role.

References

[1] Apaloo, J. (1997) Revisiting strategic models of evolution: The concept of neighborhood invader strategies. *Theor. Pop. Biol.* **52**, 71-77.

[2] Apaloo, J., J.S. Brown and T.L. Vincent (2009) Evolutionary game theory: ESS, convergence stability, and NIS. *Evol. Ecol. Res.* **11**, 489-515.

[3] Bomze, I.M. (1991) Cross entropy minimization in uninvadable states of complex populations. *J. Math. Biol.* **30**, 73-87.

[4] Bomze, I.M. and B.M. Pötscher (1989) *Game Theoretical Foundations of Evolutionary Stability.* Springer, Berlin.

[5] Chamberland, M. and R. Cressman (2000) An example of dynamic (in)consistency in symmetric extensive form evolutionary games. *Games and Econ. Behav.* **30**, 319-326.

[6] Christiansen, F.B. (1991) On conditions for evolutionary stability for a continuously varying character. *Amer. Nat.* **138**, 37-50.

[7] Courteau, J. and S. Lessard (2000) Optimal sex ratios in structured populations. *J. Theor. Biol.*, **207**, 159-175.

[8] Cressman, R. (2003) *Evolutionary Dynamics and Extensive Form Games.* MIT Press, Cambridge, MA.

[9] Cressman, R. (2009) Continuously stable strategies, neighborhood superiority and two-player games with continuous strategy spaces. *Int. J. Game Theory* **38**, 221-247.

[10] Cressman, R. (2010) CSS, NIS and dynamic stability for two-species behavioral models with continuous trait spaces. *J. Theor. Biol.* **262**, 80-89.

[11] Cressman, R., J. Garay and J. Hofbauer (2001) Evolutionary stability concepts for N-species frequency-dependent interactions. *J. Theor. Biol.* **211**, 1-10.

[12] Cressman, R., J. Hofbauer and F. Riedel (2006) Stability of the replicator equation for a single-species with a multi-dimensional continuous trait space. *J. Theor. Biol.* **239**, 273-288.

[13] Dercole, F. and S. Rinaldi (2008) *Analysis of Evolutionary Processes. The Adaptive Dynamics Approach and its Applications.* Princeton University Press, Princeton.

[14] Dieckmann, U. and R. Law (1996) The dynamical theory of coevolution: a derivation from stochastic ecological processes. *J. Math. Biol.* **34**, 579-612.

[15] Doebeli, M. and U. Dieckmann (2000) Evolutionary branching and sympatric speciation caused by different types of ecological interactions. *Am. Nat.* **156**, S77-S101.

[16] Eshel, I. (1983) Evolutionary and continuous stability. *J. Theor. Biol.* **103**, 99-111.

[17] Fretwell, D.S. and H.L. Lucas (1969) On territorial behavior and other factors influencing habitat distribution in birds. *Acta Biotheoretica* **19**, 16-32.

[18] Geritz, S.A.H., É. Kisdi, G. Meszéna and J.A.J. Metz (1998) Evolutionarily singular strategies and the adaptive growth and branching of the evolutionary tree. *Evol. Ecol.* **12**, 35-57.

[19] Hofbauer, J. and K. Sigmund (1998) *Evolutionary Games and Population Dynamics.* Cambridge University Press, Cambridge.

[20] Kisdi, É. and G. Meszéna (1995) Life histories with lottery competition in a stochastic environment: ESSs which do not prevail. *Theor. Pop. Biol.* **47**, 191-211.

[21] Krivan, V., R. Cressman and C. Schneider (2008) The ideal free distribution: A review and synthesis of the game theoretic perspective. *Theor. Pop. Biol.* **73**, 403-425.

[22] Kuhn, H. (1953) Extensive games and the problem of information. In H. Kuhn and A. Tucker, eds., *Contributions to the Theory of Games II. Annals of Mathematics 28.* Princeton University Press, Princeton.

[23] Leimar, O. (2009). Multidimensional convergence stability. *Evol. Ecol. Res.* **11**, 191-208..

[24] Lessard, S. (1990) Evolutionary stability: one concept, several meanings. *Theor. Pop. Biol.* **37**, 159-170.

[25] Marrow P., U. Dieckmann and R. Law (1996) Evolutionary dynamics of predator-prey systems: an ecological perspective. *J. Math. Biol.* **34**, 556-578.

[26] Maynard Smith, J. (1982) *Evolution and the Theory of Games.* Cambridge University Press, Cambridge.

[27] Oechssler, J. and F. Riedel (2001) Evolutionary dynamics on infinite strategy spaces. *Econ. Theory* **17**, 141-162.

[28] Rosenthal, R. (1981) Games of perfect information, predatory pricing and the chain-store paradox. *J. Econ. Theor.* **25**, 92-100.

[29] Selten, R. (1978) The chain-store paradox. *Theory and Decision* **9**, 127-159.

[30] Selten, R. (1983) Evolutionary stability in extensive two-person games. *Math. Soc. Sci.* **5**, 269-363.
[31] Selten, R. (1988) Evolutionary stability in extensive two-person games - correction and further development. *Math. Soc. Sci.* **16**, 223-266.
[32] van Damme, E. (1991) *Stability and Perfection of Nash Equilibria* (2nd Edition). Springer-Verlag, Berlin.
[33] Vincent, T.L. and J.S. Brown (2005) *Evolutionary Game Theory, Natural Selection and Darwinian Dynamics*, Cambridge University Press.
[34] von Neumann, J. and O. Morgenstern (1944) *Theory of Games and Economic Behavior.* Princeton University Press, Princeton.
[35] Weibull, J. (1995) *Evolutionary Game Theory.* MIT Press, Cambridge, MA.

DEPARTMENT OF MATHEMATICS, WILFRID LAURIER UNIVERSITY WATERLOO, ONTARIO N2L 3C5 CANADA
E-mail address: rcressman@wlu.ca

Proceedings of Symposia in Applied Mathematics
Volume **69**, 2011

Deterministic Evolutionary Game Dynamics

Josef Hofbauer

ABSTRACT. This is a survey about continuous time deterministic evolutionary dynamics for finite games. In particular, six basic dynamics are described: the replicator dynamics, the best response dynamics, the Brown–von Neumann–Nash dynamics, the Smith dynamics, and the payoff projection dynamics. Special classes of games, such as stable games, supermodular games and partnership games are discussed. Finally a general nonconvergence result is presented.

1. Introduction: Evolutionary Games

We consider a large population of players, with a finite set of pure strategies $\{1, \dots, n\}$. x_i denotes the frequency of strategy i. $\Delta_n = \{x \in \mathbb{R}^n : x_i \geq 0, \sum_{i=1}^n x_i = 1\}$ is the $(n-1)$–dimensional simplex which will often be denoted by Δ if there is no confusion.

The payoff to strategy i in a population x is $a_i(x)$, with $a_i : \Delta \to \mathbb{R}$ a continuous function (population game). The most important special case is that of a symmetric two person game with $n \times n$ payoff matrix $A = (a_{ij})$; with random matching this leads to the linear payoff function $a_i(x) = \sum_j a_{ij}x_j = (Ax)_i$.

$\hat{x} \in \Delta$ is a Nash equilibrium (NE) iff

$$\hat{x} \cdot a(\hat{x}) \geq x \cdot a(\hat{x}) \quad \forall x \in \Delta. \tag{1.1}$$

Occasionally I will also look at bimatrix games (played between two player populations), with $n \times m$ payoff matrices A, B, or at N person games.

Evolutionarily stable strategies. According to Maynard Smith [**36**], a mixed strategy $\hat{x} \in \Delta$ is an *evolutionarily stable strategy* (ESS) if

$$\begin{array}{llcll} \text{(i)} & x \cdot A\hat{x} & \leq & \hat{x} \cdot A\hat{x} & \forall x \in \Delta, \quad \text{and} \\ \text{(ii)} & x \cdot Ax & < & \hat{x} \cdot Ax & \text{for } x \neq \hat{x}, \text{ if there is equality in (i).} \end{array}$$

The first condition (i) is simply Nash's definition (1.1) for an equilibrium. It is easy to see that $\hat{x}$ is an ESS, iff $\hat{x} \cdot Ax > x \cdot Ax$ holds for all $x \neq \hat{x}$ in a neighbourhood of $\hat{x}$. This property is called *locally superior* in [**61**]. For an interior equilibrium

2000 *Mathematics Subject Classification.* Primary 91A22.
I thank Karl Sigmund for comments, and Bill Sandholm and Francisco Franchetti for producing the figures with their *Dynamo* software [**50**].

$\hat{x}$, the equilibrium condition $\hat{x}\cdot A\hat{x} = x\cdot A\hat{x}$ for all $x \in \Delta$ together with (ii) implies $(\hat{x} - x)\cdot A(x - \hat{x}) > 0$ for all x and hence

$$z\cdot Az < 0 \qquad \forall z \in \mathbb{R}^n_0 = \{z \in \mathbb{R}^n : \sum_i z_i = 0\} \quad \text{with} \quad z \neq 0. \tag{1.2}$$

Condition (1.2) says that the mean payoff $x\cdot Ax$ is a strictly concave function on Δ. Conversely, games satisfying (1.2) have a unique ESS (possibly on the boundary) which is also the unique Nash equilibrium of the game. The slightly weaker condition

$$z\cdot Az \leq 0 \qquad \forall z \in \mathbb{R}^n_0 \tag{1.3}$$

includes also the limit cases of zero–sum games and games with an interior equilibrium that is a 'neutrally stable' strategy (i.e., equality is allowed in (ii)). Games satisfying (1.3) need no longer have a unique equilibrium, but the set of equilibria is still a nonempty convex subset of Δ.

For the *rock–scissors–paper* game with (a cyclic symmetric) pay-off matrix

$$A = \begin{pmatrix} 0 & -b & a \\ a & 0 & -b \\ -b & a & 0 \end{pmatrix} \qquad \text{with } a, b > 0 \tag{1.4}$$

with the unique Nash equilibrium $E = (\frac{1}{3}, \frac{1}{3}, \frac{1}{3})$ we obtain the following: for $z \in \mathbb{R}^3_0$, $z_1 + z_2 + z_3 = 0$,

$$z\cdot Az = (a - b)(z_1z_2 + z_2z_3 + z_1z_3) = \frac{b-a}{2}[z_1^2 + z_2^2 + z_3^2].$$

Hence for $0 < b < a$, the game is negative definite, and E is an ESS. On the other hand, if $0 < a < b$, the game is positive definite:

$$z\cdot Az > 0 \quad \forall z \in \mathbb{R}^n_0 \setminus \{0\}, \tag{1.5}$$

the equilibrium E is not evolutionarily stable, indeed the opposite, and might be called an '*anti–ESS*'.

For a classical game theorist, all RPS games are the same. There is a unique Nash equilibrium, even a unique correlated equilibrium [**60**], for any $a, b > 0$. In evolutionary game theory the dichotomy $a < b$ versus $a > b$ is crucial, as we will see in the next sections, in particular in the figures 1–6.

2. Game Dynamics

In this section I present 6 special (families of) game dynamics. As we will see they enjoy a particularly nice property: Interior ESS are globally asymptotically stable.

The presentation follows largely [**22, 24, 28**].

1. Replicator dynamics
2. Best response dynamics
3. Logit dynamics (and other smoothed best reply dynamics)
4. Brown–von Neumann–Nash dynamics
5. Payoff comparison dynamics
6. Payoff projection dynamics

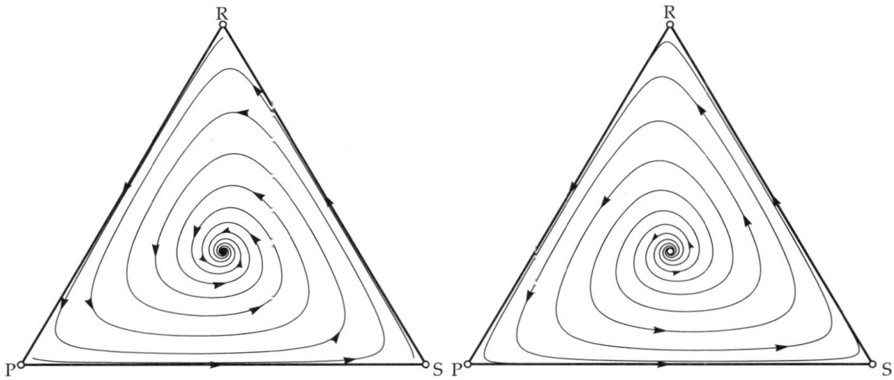

FIGURE 1. Replicator dynamics for Rock-Paper-Scissors games: $a > b$ versus $a < b$

Replicator dynamics.

$$\dot{x}_i = x_i\left(a_i(x) - x\cdot a(x)\right), \quad i = 1, \dots, n \qquad \text{(REP)} \tag{2.1}$$

In the zero-sum version $a = b$ of the RSP game, all interior orbits are closed, circling around the interior equilibrium E, with $x_1x_2x_3$ as a constant of motion.

THEOREM 2.1. *In a negative definite game satisfying* (1.2), *the unique Nash equilibrium $p \in \Delta$ is globally asymptotically stable for (REP). In particular, an interior ESS is globally asymptotically stable.*
On the other hand, in a positive definite game satisfying (1.5) *with an interior equilibrium p, i.e., an anti-ESS, p is a global repellor. All orbits except p converge to the boundary $bd\,\Delta$.*

The proof uses $V(x) = \prod x_i^{p_i}$ as a Lyapunov function.
For this and further results on (REP) see Sigmund's chapter [**53**], and [**9, 26, 27, 48, 61**].

Best response dynamics. In the best response dynamics[1] [**14, 35, 19**] one assumes that in a large population, a small fraction of the players revise their strategy, choosing best replies[2] $BR(x)$ to the current population distribution x.

$$\dot{x} \in BR(x) - x. \tag{2.2}$$

Since best replies are in general not unique, this is a differential *inclusion* rather than a differential equation. For continuous payoff functions $a_i(x)$ the right hand side is a non-empty convex, compact subset of Δ which is upper semi-continuous in x. Hence solutions exist, and they are Lipschitz functions $x(t)$ satisfying (2.2) for almost all $t \geq 0$, see [1].

For games with linear payoff, solutions can be explicitely constructed as piecewise linear functions, see [**9, 19, 27, 53**].

For interior NE of linear games we have the following stability result [**19**].

[1]For bimatrix games, this dynamics is closely related to the 'fictitious play' by Brown [**6**], see Sorin's chapter [**56**].

[2]Recall the set of best replies $\mathrm{BR}(x) = \mathrm{Argmax}_{y\in\Delta}\ y\cdot a(x) = \{y \in \Delta : y\cdot a(x) \geq z\cdot a(x) \forall z \in \Delta\} \subseteq \Delta$.

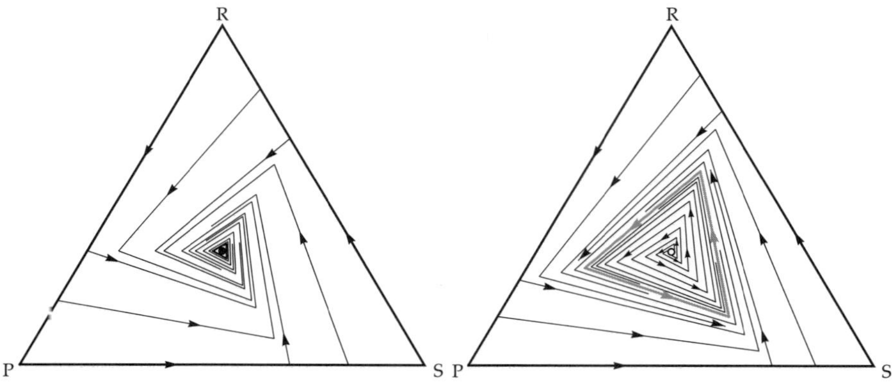

FIGURE 2. Best response dynamics for Rock-Paper-Scissors games, left one: $a \geq b$, right one: $a < b$

Let $\mathfrak{B} = \{b \in \mathrm{bd}\Delta_n : (Ab)_i = (Ab)_j \text{ for all } i,j \in \mathrm{supp}(b)\}$ denote the set of all rest points of (REP) on the boundary. Then the function[3]

$$w(x) = \max\left\{\sum_{b\in\mathfrak{B}} b\cdot Ab\, u(b) : u(b) \geq 0, \sum_{b\in\mathfrak{B}} u(b) = 1, \sum_{b\in\mathfrak{B}} u(b)b = x\right\} \tag{2.3}$$

can be interpreted in the following way. Imagine the population in state x being decomposed into subpopulations of size $u(b)$ which are in states $b \in \mathfrak{B}$, and call this a *$\mathfrak{B}$–segregation of b*. Then $w(x)$ is the maximum mean payoff the population x can obtain by such a $\mathfrak{B}$–segregation. It is the smallest concave function satisfying $w(b) \geq b\cdot Ab$ for all $b \in \mathfrak{B}$.

THEOREM 2.2. *The following three conditions are equivalent:*
(a) There is a vector $p \in \Delta_n$, such that $p\cdot Ab > b\cdot Ab$ holds for all $b \in \mathfrak{B}$.
(b) $V(x) = \max_i(Ax)_i - w(x) > 0$ for all $x \in \Delta_n$.
(c) There exist a unique interior equilibrium $\hat{x}$, and $\hat{x}\cdot A\hat{x} > w(\hat{x})$.
These conditions imply:
The equilibrium $\hat{x}$ is reached in finite and bounded time by any BR path.

The proof consists in showing that the function V from (b) decreases along the solutions of the BR dynamics (2.2).

In the rock–scissors–paper game the set $\mathfrak{B}$ reduces to the set of pure strategies, the Lyapunov function is simply $V(x) = \max_i(Ax)_i$ and satisfies $\dot{V} = -V$ (except at the NE), see [**13**]. Since $\min_{x\in\Delta} V(x) = V(\hat{x}) = \hat{x}\cdot A\hat{x} = \frac{a-b}{3} > 0$ the exponentially decreasing $V(x(t))$ reaches this minimum value after a finite time. So all orbits reach the NE in finite time.

If $p \in \mathrm{int}\,\Delta$ is an interior ESS then condition (a) holds not only for all $b \in \mathfrak{B}$ but for all $b \neq p$. In this case the Lyapunov function $V(x) = \max_i(Ax)_i - x\cdot Ax \geq 0$ can also be used. This leads to

THEOREM 2.3. [**22**] *For a negative semidefinite game* (1.3) *the convex set of its equilibria is globally asymptotically stable for the best–response dynamics.*[4]

[3]If $\mathfrak{B}$ is infinite it is sufficient to take the finitely many extreme points of its convex pieces.

[4]Using the tools in Sorin's chapter [**56**, section 1] this implies also global convergence of (discrete time) fictitious play. A similar result holds also for nonlinear payoff functions, see [**24**].

PROOF. The Lyapunov function $V(x) = \max_i (Ax)_i - x \cdot Ax \geq 0$ satisfies $\dot{V} = \dot{x} \cdot A\dot{x} - \dot{x} \cdot Ax < 0$ along piecewise linear solutions outside the set of NE. □

Note that for zero–sum games, $\dot{V} = -V$, so $V(x(t)) = e^{-t}V(x(0)) \to 0$ as $t \to \infty$, so $x(t)$ converges to the set of NE. For negative definite games, $\dot{V} < -c - V$ for some $c > 0$ and hence $x(t)$ reaches the NE in finite time.

For positive definite RSP games ($b > a$), $V(x) = \max_i (Ax)_i$ still satisfies $\dot{V} = -V$. Hence the NE is a repeller and all orbits (except the constant one at the NE) converge to the set where $V(x) = \max_i (Ax)_i = 0$ which is a closed orbit under the BR dynamics. It is called the *Shapley triangle* of the game, as a tribute to [**52**], see figure 2 (right). In this case the equilibrium payoff $\frac{a-b}{3}$ is smaller than 0, the payoff for a tie. This is the intuitive reason why the population tries to get away from the NE and closer to the pure states.

Interestingly, the times averages of the solutions of the replicator dynamics approach for $b > a$ the very same Shapley triangle, see [**13**]. The general reason for this is explained in Sorin's chapter [**56**, ch. 3].

For similar cyclic games with $n = 4$ strategies several Shapley polygons can coexist, see [**16**]. For $n \geq 5$ chaotic dynamics is likely to occur.

Smoothed best replies. The BR dynamics can be approximated by smooth dynamics such as the *logit dynamics* [**5, 12, 31**]

$$\dot{x} = L(\frac{Ax}{\varepsilon}) - x \tag{2.4}$$

with

$$L : \mathbb{R}^n \to \Delta, \quad L_k(u) = \frac{e^{u_k}}{\sum_j e^{u_j}}.$$

with $\varepsilon > 0$. As $\varepsilon \to 0$, this approaches the best reply dynamics, and every family of rest points[5] $\hat{x}_\varepsilon$ accumulates in the set of Nash equilibria.

There are (at least) two ways to motivate and generalize this 'smoothing'.

Whereas $BR(x)$ is the set of maximizers of the linear function $z \mapsto \sum_i z_i a_i(x)$ on Δ, consider $b_{\varepsilon v}(x)$, the unique maximizer of the function $z \mapsto \sum_i z_i a_i(x) + \varepsilon v(z)$ on int Δ, where $v : \text{int}\, \Delta \to \mathbb{R}$ is a strictly concave function such that $|v'(z)| \to \infty$ as z approaches the boundary of Δ. If v is the entropy $-\sum z_i \log z_i$, the corresponding smoothed best reply dynamics

$$\dot{x} = b_{\varepsilon v}(x) - x \tag{2.5}$$

reduces to the logit dynamics (2.4) above [**12**]. Another choice[6] is $v(x) = \sum_i \log x_i$ used by Harsányi [**17**] in his logarithmic games.

Another way to perturb best replies are stochastic perturbations. Let ε be a random vector in $\mathbb{R}^n$ distributed according to some positive density function. For $z \in \mathbb{R}^n$, let

$$C_i(z) = \text{Prob}(z_i + \varepsilon_i \geq z_j + \varepsilon_j \quad \forall j), \tag{2.6}$$

and $b(x) = C(a(x))$ the resulting stochastically perturbed best reply function. It can be shown [**23**] that each such stochastic perturbation can be represented by a deterministic perturbation as described before. The main idea is that there is a

[5] These are the quantal response equilibria of McKelvey and Palfrey [**37**].

[6] A completely different approximate best reply function appears already in Nash's Ph.D. thesis [**40**], in his first proof of the existence of equilibria by Brouwer's fixed point theorem.

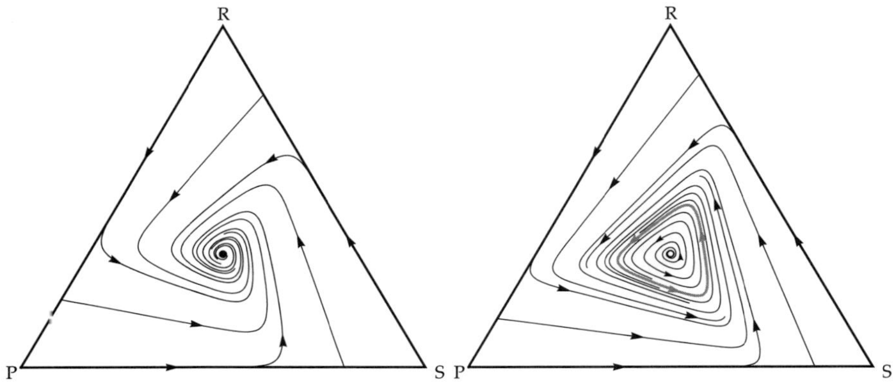

FIGURE 3. Logit dynamics for Rock-Paper-Scissors games: $a \geq b$ versus $a < b$

potential function $W : \mathbb{R}^n \to \mathbb{R}$, with $\frac{\partial W}{\partial a_i} = C_i(a)$ which is convex, and has $-v$ as its Legendre transform. If the (ε_i) are i.i.d. with the extreme value distribution $F(x) = \exp(-\exp(-x))$ then $C(a) = L(a)$ is the logit choice function and we obtain (2.4).

THEOREM 2.4. [**22**] *In a negative semidefinite game* (1.3), *the smoothed BR dynamics* (2.5) *(including the logit dynamics) has a unique equilibrium* $\hat{x}_\varepsilon$. *It is globally asymptotically stable.*

The proof uses the Lyapunov function

$$V(x) = \pi_{\varepsilon v}(b_{\varepsilon v}(x), x) - \pi_{\varepsilon v}(x, x) \geq 0 \quad \text{with} \quad \pi_{\varepsilon v}(z, x) = z \cdot a(x) + \varepsilon v(z).$$

I will return to these perturbed dynamics in section 5. For more information on the logit dynamics see Sorin's chapter [**56**] and references therein, and [**43**].

The Brown–von Neumann–Nash dynamics. The *Brown–von Neumann–Nash dynamics* (BNN) is defined as

$$\dot{x}_i = \hat{a}_i(x) - x_i \sum_{j=1}^{n} \hat{a}_j(x), \tag{2.7}$$

where

$$\hat{a}_i(x) = [a_i(x) - x \cdot a(x)]_+ \tag{2.8}$$

(with $u_+ = \max(u, 0)$) denotes the positive part of the excess payoff for strategy i. This dynamics is closely related to the continuous map $f : \Delta \to \Delta$ defined by

$$f_i(x) = \frac{x_i + h\hat{a}_i(x)}{1 + h\sum_{j=1}^{n} \hat{a}_j(x)} \tag{2.9}$$

which Nash [**41**] used (for $h = 1$) to prove the existence of equilibria, by applying Brouwer's fixed point theorem: It is easy to see that $\hat{x}$ is a fixed point of f iff it is a rest point of (2.7) iff $\hat{a}_i(\hat{x}) = 0$ for all i, i.e. iff $\hat{x}$ is a Nash equilibrium of the game.

Rewriting the Nash map (2.9) as a difference equation, and taking the limit $\lim_{h\to 0} \frac{f(x)-x}{h}$ yields (2.7). This differential equation was considered earlier by

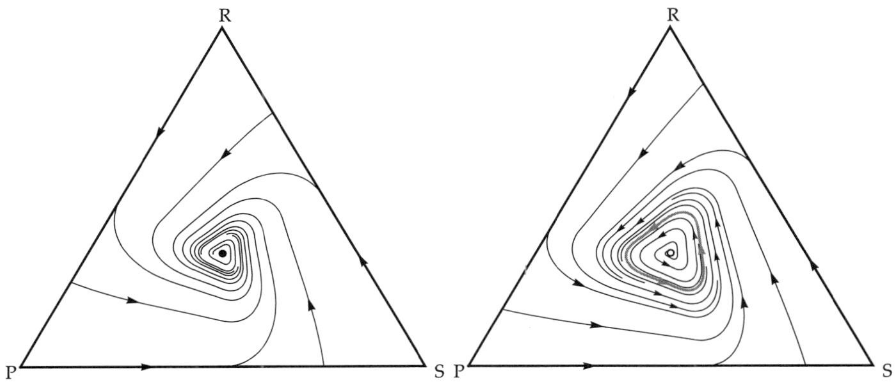

FIGURE 4. BNN dynamics for Rock-Paper-Scissors games: $a \geq b$ versus $a < b$

Brown and von Neumann [**7**] in the special case of zero–sum games, for which they proved global convergence to the set of equilibria.

In contrast to the best reply dynamics, the BNN dynamics (2.7) is Lipschitz (if payoffs are Lipschitz) and hence has unique solutions.

Equation (2.7) defines an 'innovative better reply' dynamics. A strategy not present that is a best (or at least a better) reply against the current population will enter the population.

THEOREM 2.5. [**7, 22, 24, 42**] *For a negative semidefinite game* (1.3)*, the convex set of its equilibria is globally asymptotically stable for the BNN dynamics* (2.7).

The proof uses the Lyapunov function $V = \frac{1}{2}\sum_i \hat{a}_i(x)^2$, since $V(x) \geq 0$ with equality at NE, and

$$\dot{V} = \dot{x}\cdot A\dot{x} - \dot{x}\cdot Ax \sum_i \hat{a}_i(x) \leq 0,$$

with equality ony at NE.

Dynamics based on pairwise comparison. The BNN dynamics is a prototype of an innovative dynamics. A more natural way to derive innovative dynamics is the following,

$$\dot{x}_i = \sum_j x_j \rho_{ji} - x_i \sum_j \rho_{ij}, \tag{2.10}$$

in the form of an input–output dynamics. Here $x_i\rho_{ij}$ is the flux from strategy i to strategy j, and $\rho_{ij} = \rho_{ij}(x) \geq 0$ is the rate at which an i player switches to the j strategy.

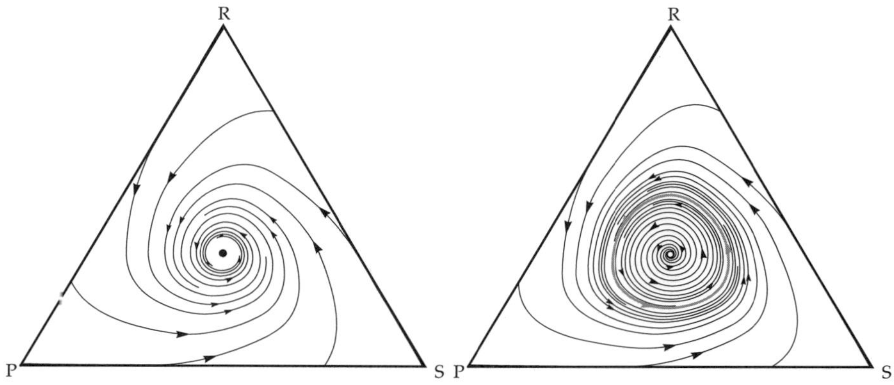

FIGURE 5. Smith's pairwise difference dynamics for Rock-Paper-Scissors games: $a \geq b$ versus $a < b$

A natural assumption on the *revision protocol*[7] ρ is

$$\rho_{ij} > 0 \Leftrightarrow a_j > a_i, \quad \text{and} \quad \rho_{ij} \geq 0.$$

Here switching to *any* better reply is possible, as opposed to the BR dynamics where switching is only to the optimal strategies (usually there is only one of them), or the BNN dynamics where switching occurs only to strategies better than the population average.

An important special case is when the switching rate depends on the payoff difference only, i.e.,

$$\rho_{ij} = \phi(a_j - a_i) \tag{2.11}$$

where ϕ is a function with $\phi(u) > 0$ for $u > 0$ and $\phi(u) = 0$ for $u \leq 0$. The resulting dynamics (2.10) is called *pairwise comparison dynamics.* The natural choice seems $\phi(u) = u_+$, given by the proportional rule

$$\rho_{ij} = [a_j - a_i]_+. \tag{2.12}$$

The resulting *pairwise difference dynamics* (PD)

$$\dot{x}_i = \sum_j x_j [a_i - a_j]_+ - x_i \sum_j [a_j - a_i]_+ \tag{2.13}$$

was introduced by Michael J. Smith [**55**] in the transportation literature as a dynamic model for congestion games. He also proved the following global stability result.

THEOREM 2.6. [**55**] *For a negative semidefinite game* (1.3)*, the convex set of its equilibria is globally asymptotically stable for the PD dynamics* (2.13).

[7]All the basic dynamics considered so far can be written in the form (2.10) with a suitable revision protocol ρ (with some obvious modification in the case of the multi–valued BR dynamics). Given the revision protocol ρ, the payoff function a, and a finite population size N, there is a natural finite population model in terms of a Markov process on the grid $\{x \in \Delta : Nx \in \mathbb{Z}^n\}$. The differential equation (2.10) provides a very good approximation of the behavior of this stochastic process, at least over finite time horizons and for large population sizes. For all this see Sandholm's chapter [**49**].

The proof uses the Lyapunov function $V(x) = \sum_{i,j} x_j[a_i(x) - a_j(x)]_+^2$, by showing $V(x) \geq 0$ and $V(x) = 0$ iff x is a NE, and

$$2\dot{V} = \dot{x}\cdot A\dot{x} + \sum_{k,j} x_k \rho_{kj} \sum_i (\rho_{ji}^2 - \rho_{ki}^2) < 0$$

except at NE. This result extends to pairwise comparison dynamics (2.10,2.11), see [**24**].

The payoff projection dynamics. A more recent proof of the existence of Nash equilibria, due to *Gül–Pearce–Stacchetti* [**15**] uses the payoff projection map

$$P_h x = \Pi_\Delta(x + ha(x))$$

Here $h > 0$ is fixed and $\Pi_\Delta : \mathbb{R}^n \to \Delta$ is the projection onto the simplex Δ, assigning to each vector $u \in \mathbb{R}^n$ the point in the compact convex set Δ which is closest to u. Now $\Pi_\Delta(z) = y$ iff for all $x \in \Delta$, the angle between $x - y$ and $z - y$ is obtuse, i.e., iff $(x-y)\cdot(z-y) \leq 0$ for all $x \in \Delta$. Hence, $P_h\hat{x} = \hat{x}$ iff for all $x \in \Delta$, $(x - \hat{x})\cdot a(\hat{x}) \leq 0$, i.e., iff $\hat{x}$ is a Nash equilibrium. Since the map $P_h : \Delta \to \Delta$ is continuous Brouwer's fixed point theorem implies the existence of a Nash equilibrium.

Writing this map as a difference equation, we obtain in the limit $h \to 0$

$$\dot{x} = \lim_{h\to 0} \frac{\Pi_\Delta(x + ha(x)) - x}{h} = \Pi_{T(x)} a(x) \tag{2.14}$$

with

$$T(x) = \{\xi \in \mathbb{R}^n : \sum_i \xi_i = 0, \xi_i \geq 0 \text{ if } x_i = 0\}$$

being the cone of feasible directions at x into Δ.

This is the *payoff projection dynamics* (PP) of Lahkar and Sandholm [**34**]. The latter equality in (2.14) and its dynamic analysis use some amount of convex analysis, in particular the Moreau decomposition, see [**1, 34**].

For $x \in \operatorname{int}\Delta$ we obtain

$$\dot{x}_i = a_i(x) - \frac{1}{n}\sum_k a_k(x)$$

which, for a linear game, is simply a linear dynamics. It appeared in many places as a suggestion for a simple game dynamics, but how to treat this on the boundary has been rarely dealt with. Indeed, the vector field (2.14) is discontinuous on $\operatorname{bd}\Delta$. However, essentially because P_h is Lipschitz, solutions exist for all $t \geq 0$ and are unique (in forward time). This can be shown by rewriting (2.14) as a viability problem in terms of the normal cone ([**1, 34**])

$$\dot{x} \in a(x) - N_\Delta(x), \quad x(t) \in \Delta.$$

THEOREM 2.7. [**34**] *In a negative definite game* (1.2), *the unique NE is globally asymptotically stable for the payoff projection dynamics* (2.14).

The proof uses as Lyapunov function the Euclidean distance to the equilibrium $V(x) = \sum_i (x_i - \hat{x}_i)^2$.

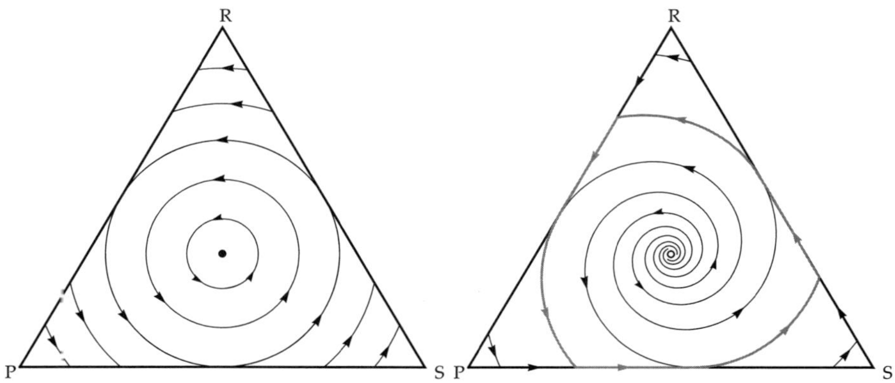

FIGURE 6. Payoff projection dynamics for Rock-Paper-Scissors games: $a = b$ versus $a < b$

Summary. As we have seen many of the special dynamics are related to maps that have been used to prove existence of Nash equilibria. The best response dynamics, the perturbed best response dynamics, and the BNN dynamics correspond to the three proofs given by Nash himself: [**39, 40, 41**]. The payoff projection dynamics is related to [**15**]. Even the replicator dynamics can be used to provide such a proof, if only after adding a mutation term, see [**26, 27**], or Sigmund's chapter [**53**, (11.3)]:

$$\dot{x}_i = x_i \left(a_i(x) - x \cdot a(x) \right) + \varepsilon_i - x_i \sum_j \varepsilon_j, \quad i = 1, \dots, n \tag{2.15}$$

with $\varepsilon_i > 0$ describing mutation rates.

Moreover, there is a result analogous to Theorem 2.4.

THEOREM 2.8. *For a negative semidefinite game* (1.3), *and any* $\varepsilon_i > 0$, (2.15) *has a unique rest point* $\hat{x}(\varepsilon) \in \Delta$. *It is globally asymptotically stable, and for* $\varepsilon \to 0$ *it approaches the set of NE of the game.*

I show a slightly more general result. With the notation $\phi_i(x_i) = \frac{\varepsilon_i}{x_i}$, let us rewrite (2.15) as

$$\dot{x}_i = x_i \left((Ax)_i + \phi_i(x_i) - x \cdot Ax - \bar{\phi} \right) \tag{2.16}$$

where $\bar{\phi} = \sum_i x_i \phi_i(x_i)$. In the following, I require only that each ϕ_i is a strictly decreasing function.

THEOREM 2.9. *If A is a negative semidefinite game* (1.3) *and the functions ϕ_i are strictly decreasing, then there is a unique rest point $\hat{x}(\phi)$ for* (2.16) *which is globally asymptotically stable for* (2.16).

PROOF. By Brouwer's fixed point theorem, (2.16) has a rest point $\hat{x} \in \Delta$. Consider now the function $L(x) = \sum_i \hat{x}_i \log x_i$ defined on $\operatorname{int}\Delta$. Then

$$\begin{aligned} \dot{L} &= \sum_i \hat{x}_i \frac{\dot{x}_i}{x_i} = \sum_i \hat{x}_i \left((Ax)_i + \phi_i(x_i) - x \cdot Ax - \bar{\phi}\right) \\ &= \sum_i (\hat{x}_i - x_i)\left((Ax)_i - \phi_i(x_i)\right) \\ &= -(\hat{x} - x) \cdot A(\hat{x} - x) - \sum_i (x_i - \hat{x}_i)\left(\phi_i(\hat{x}_i) - \phi_i(x_i)\right) \geq 0 \end{aligned}$$

with equality only at $x = \hat{x}$. Hence L is a Lyapunov function for $\hat{x}$, and hence $\hat{x}$ is globally asymptotically stable (w.r.t. $\operatorname{int}\Delta$). □

The six basic dynamics described so far enjoy the following common properties.

1. The unique NE of a negative definite game (in particular, any interior ESS) is globally asymptotically stable.
2. Interior NE of a positive definite game ('anti-ESS') are repellors.

Because of the nice behaviour of negative (semi-)definite games with respect to these basic dynamics, Sandholm christened them **stable games**.

For nonlinear games in a single population these are games whose payoff function $a : \Delta \to \mathbb{R}^n$ satisfies

$$(a(x) - a(y))(x - y) \leq 0 \qquad \forall x, y \in \Delta \tag{2.17}$$

or equivalently, if a is smooth,

$$z \cdot a'(x) z \leq 0 \qquad \forall x \in \Delta, z \in \mathbb{R}^n_0$$

Examples are congestion games [**48**], the war of attrition [**36**], the sex–ratio game [**36**], the habitat selection game [**10**], or simply the nonlinear payoff function $a(x) = Ax + \phi(x)$ in (2.16).

The global stability theorems 2.1, 2.3, 2.4, 2.5, 2.6, 2.14 hold for general stable N population games, see [**24, 48**].

3. Bimatrix games

The replicator dynamics for an $n \times m$ bimatrix game (A, B) reads

$$\begin{aligned} \dot{x}_i &= x_i\big((Ay)_i - x \cdot Ay\big), \quad i = 1, \ldots, n \\ \dot{y}_j &= y_j\big((B^T x)_j - x \cdot By\big) \quad j = 1, \ldots, m \end{aligned}$$

For its properties see [**26, 27**] and especially [**21**]. N person games are treated in [**61**] and [**44**].

The best reply dynamics for bimatrix games reads

$$\dot{x} \in BR^1(y) - x \qquad \dot{y} \in BR^2(x) - y \tag{3.1}$$

See Sorin [**56**, section 1] for more information.

For 2×2 games the state space $[0, 1]^2$ is two-dimensional and one can completely classify the dynamic behaviour. There are four robust cases for the replicator dynamics, see [**26, 27**], and additionally 11 degenerate cases. Some of these degenerate cases arise naturally as extensive form games, such as the Entry Deterrence

Game, see Cressman's chapter [**10**]. A complete analysis including all phase portraits are presented in [**9**] for (BR) and (REP), and in [**46**] for the BNN and the Smith dynamics.

For bimatrix games, stable games include zero-sum games, but not much more. We call an $n \times m$ bimatrix game (A, B) a *rescaled zero-sum game* [**26, 27**] if

$$\exists c > 0 : \quad u \cdot Av = -cu \cdot Bv \qquad \forall u \in \mathbb{R}_0^n, v \in \mathbb{R}_0^m \tag{3.2}$$

or equivalently, there exists an $n \times m$ matrix C, $\alpha_i, \beta_j \in \mathbb{R}$ and $\gamma > 0$ s.t.

$$a_{ij} = c_{ij} + \alpha_j, \quad b_{ij} = -\gamma c_{ij} + \beta_i, \quad \forall i = 1, \ldots, n, j = 1, \ldots, m$$

For 2×2 games, this includes an open set of payoff matrices, corresponding to games with a cyclic best reply structure, or equivalently, those with a unique and interior Nash equilibrium. Simple examples are the *Odd or Even* game [**53**, (1.1)], or the Buyers and Sellers game [**10**]. However, for larger n, m this is a thin set of games, e.g. for 3×3 games, this set has codimension 3.

For such rescaled zero-sum games, the set of Nash equilibria is stable for (REP), (BR) and the other basic dynamics.

One of the main open problems in evolutionary game dynamics concerns the converse.

CONJECTURE 3.1. *Let (p, q) be an isolated interior equilibrium of a bimatrix game (A, B), which is stable for the BR dynamics or for the replicator dynamics. Then $n = m$ and (A, B) is a rescaled zero sum game.*

4. Dominated Strategies

A pure strategy i (in a single population game with payoff function $a : \Delta \to \mathbb{R}^n$) is said to be *strictly dominated* if there exists some $y \in \Delta$ such that

$$a_i(x) < y \cdot a(x) \tag{4.1}$$

for all $x \in \Delta$. A rational player will not use such a strategy.

In the best response dynamics, $\dot{x}_i = -x_i$ and hence $x_i(t) \to 0$ as $t \to \infty$.

Similarly, for the replicator dynamics, $L(x) = \log x_i - \sum_k y_k \log x_k$ satisfies $\dot{L}(x) < 0$ for $x \in \operatorname{int} \Delta$ and hence $x_i(t) \to 0$ along all interior orbits of (REP).

A similar result holds for extensions of (REP), given by differential equations of the form

$$\dot{x}_i = x_i g_i(x) \tag{4.2}$$

where the functions g_i satisfy $\sum x_i g_i(x) = 0$ on Δ. The simplex Δ and its faces are invariant. Such an equation is said to be *payoff monotonic* [**61**] if for any i, j, and $x \in \Delta$

$$g_i(x) > g_j(x) \Leftrightarrow a_i(x) > a_j(x). \tag{4.3}$$

All dynamics arising from an imitative revision protocol have this property. For such payoff monotonic dynamics, if the pure strategy i is strictly dominated by another pure strategy j, i.e., $a_i(x) < a_j(x)$ for all $x \in \Delta$ then $\frac{x_i}{x_j}$ goes monotonically to zero, and hence $x_i(t) \to 0$. However, if the dominating strategy is mixed, this need no longer be true, see [**20, 30**].

The situation is even worse for all other basic dynamics from section 2, in particular, (BNN), (PD) and (PP). As shown in [**4, 25, 34**] there are games with a pure strategy i being strictly dominated by another pure strategy j such that

i survives in the long run, i.e., $\liminf_{t\to+\infty} x_i(t) > 0$ for an open set of initial conditions.

5. Supermodular Games and Monotone Flows

An interesting class of games are the *supermodular games* (also known as games with strict strategic complementarities [**59**]). They make use of the natural order among pure strategies and are defined by

$$a_{i+1,j+1} - a_{i+1,j} - a_{i,j+1} + a_{i,j} > 0 \qquad \forall i,j \tag{5.1}$$

where $a_{i,j} = a_{ij}$ are the entries of the payoff matrix A. This means that for any $i < n$, $a_{i+1,k} - a_{ik}$ increases strictly with k.

In the case of $n = 2$ strategies this reduces to $a_{22} - a_{21} - a_{12} + a_{11} > 0$, which means that the game is positive definite (1.5). In particular, every bistable 2×2 game is supermodular.

For $n \geq 3$ there is no simple relation between supermodular games and positive definite games, although they share some properties, such as the instability of interior NE. For example, the RPS game with $b > a$ is positive definite but not supermodular. Indeed, a supermodular game cannot have a best reply cycle among the pure strategies, see below. On the other hand, an $n \times n$ pure coordination game (where the payoff matrix is a positive diagonal matrix) is positive definite, but supermodular only if $n = 2$.

Stochastic dominance defines a partial order on the simplex Δ:

$$p \succeq p' \quad \Leftrightarrow \quad \sum_{k=1}^{m} p_k \leq \sum_{k=1}^{m} p'_k \quad \forall m = 1, \ldots, n-1. \tag{5.2}$$

If all inequalities in (5.2) are strict, we write $p \succ p'$. The intuition is that p has more mass to the right than p'. This partial order extends the natural order on the pure strategies:

$1 \prec 2 \prec \cdots \prec n$. Here k is identified with the kth unit vector, i.e., a corner of Δ.

LEMMA 5.1. *Let (u_k) be an increasing sequence, and $x \preceq y$. Then $\sum u_k x_k \leq \sum u_k y_k$. If (u_k) is strictly increasing and $x \preceq y$, $x \neq y$ then $\sum u_k x_k < \sum u_k y_k$. If (u_k) is increasing but not constant and $x \prec y$ then $\sum u_k x_k < \sum u_k y_k$.*

The proof follows easily from Abel summation (the discrete analog of integration by parts): set $x_k - y_k = c_k$ and $u_{n-1} = u_n - v_{n-1}$, $u_{n-2} = u_n - v_{n-1} - v_{n-2}$, etc.

LEMMA 5.2. *For $i < j$ and $x \preceq y$, $x \neq y$:*

$$(Ax)_j - (Ax)_i < (Ay)_j - (Ay)_i. \tag{5.3}$$

PROOF. Take $u_k = a_{jk} - a_{ik}$ as strictly increasing sequence in the previous lemma. ☐

The crucial property of supermodular games is the monotonicity of the best reply correspondence.

THEOREM 5.3. [**59**] *If $x \preceq y$, $x \neq y$ then $\max BR(x) \leq \min BR(y)$, i.e., no pure best reply to y is smaller than a pure best reply to x.*

PROOF. Let $j = \max BR(x)$. Then for any $i < j$ (5.3) implies that $(Ay)_j > (Ay)_i$, hence $i \notin BR(y)$. Hence every element in $BR(y)$ is $\geq j$. □

Some further consequences of Lemma 5.2 and Theorem 5.3 are:

The extreme strategies 1 and n are either strictly dominated strategies or pure Nash equilibria.

There are no best reply cycles: Every sequence of pure strategies which is sequential best replies is finally constant and ends in a pure NE.

For results on the convergence of fictitious play and the best response dynamics in supermodular games see [**3, 32**].

THEOREM 5.4. *Mixed (=nonpure) equilibria of supermodular games are unstable under the replicator dynamics.*

PROOF. W.l.o.g., we can assume that the equilibrium $\hat{x}$ is interior (otherwise restrict to a face). A supermodular game satisfies $a_{ij} + a_{ji} < a_{ii} + a_{jj}$ for all $i \neq j$ (set $x = i, y = j$ in (5.3)). Hence, if we normalize the game by $a_{ii} = 0$, $\hat{x}\cdot A\hat{x} = \sum_{i,j}(a_{ij} + a_{ji})\hat{x}_i\hat{x}_j < 0$. Now it is shown in [**27**, p.164] that $-\hat{x}\cdot A\hat{x}$ equals the trace of the Jacobian of (REP) at $\hat{x}$, i.e., the sum of all its eigenvalues. Hence at least one of the eigenvalues has positive real part, and $\hat{x}$ is unstable. □

For different instability results of mixed equilibria see [**11**].

The following is a generalization of Theorem 5.3 to perturbed best replies, due to [**23**]. I present here a different proof.

THEOREM 5.5. *For every supermodular game*

$$x \preceq y, x \neq y \quad \Rightarrow \quad C(a(x)) \prec C(a(y))$$

holds if the choice function $C : \mathbb{R}^n \to \Delta_n$ *is* C^1 *and the partial derivatives* $C_{i,j} = \frac{\partial C_i}{\partial x_j}$ *satisfy for all* $1 \leq k, l < n$

$$\sum_{i=1}^{k}\sum_{j=1}^{l} C_{i,j} > 0, \tag{5.4}$$

and for all $1 \leq i \leq n$,

$$\sum_{j=1}^{n} C_{i,j} = 0. \tag{5.5}$$

PROOF. It is sufficient to show that the perturbed best reponse map is strongly monotone:

$$x \preceq y, x \neq y \quad \Rightarrow \quad C(Ax) \prec C(Ay)$$

From Lemma 5.2 we know: If $x \preceq y$, $x \neq y$ then $(Ay - Ax)_i$ increases strictly in i. Hence, with $a = Ax$ and $b = Ay$, it remains to show:

LEMMA 5.6. *Let* $a, b \in \mathbb{R}^n$ *with* $b_1 - a_1 < b_2 - a_2 < \cdots < b_n - a_n$. *Then* $C(a) \prec C(b)$.

This means that for each k: $C_1(a) + \cdots + C_k(a) \geq C_1(b) + \cdots + C_k(b)$.

Taking derivative in direction $u = b - a$, this follows from $\sum_{i=1}^{k}\sum_{j=1}^{n} C_{i,j}u_j < 0$ which by Lemma 5.1 holds whenever $(x_j - y_j =) c_j = \sum_{i=1}^{k} C_{i,j}$ satisfies $\sum_{j=1}^{l} c_j > 0$ for $l = 1, \ldots, n-1$ and $\sum_{j=1}^{n} c_j = 0$. □

The conditions (5.4, 5.5) on C hold for every stochastic choice model (2.6), since there $C_{i,j} < 0$ for $i \neq j$. As a consequence, the perturbed best reply dynamics

$$\dot{x} = C(a(x)) - x \tag{5.6}$$

generates a strongly monotone flow: If $x(0) \preceq y(0), x(0) \neq y(0)$ then $x(t) \prec y(t)$ for all $t > 0$. The theory of monotone flows developped by Hirsch and others (see [**54**]) implies that almost all solutions of (5.6) converge to a rest point of (5.6).

It seems that the other basic dynamics do not respect the stochastic dominance order (5.2). They do not generate a monotone flow for every supermodular game. Still there is the open problem

PROBLEM 5.7. *In a supermodular game, do almost all orbits of (BR), (REP), (BNN), (PD), (PP) converge to a NE?*

For the best reponse dynamics this entails to extend the theory of monotone flows to cover discontinuous differential equations or differential inclusions.

6. Partnership games and general adjustment dynamics

We consider now games with a symmetric payoff matrix $A = A^T$ ($a_{ij} = a_{ji}$ for all i, j). Such games are known as *partnership games* [**26, 27**] and *potential games* [**38**]. The basic population genetic model of Fisher and Haldane is equivalent to the replicator dynamics for such games, which is then a gradient system with respect to the Shahshahani metric and the mean payoff $x{\cdot}Ax$ as potential, see e.g. [**26, 27**]. The resulting increase of mean fitness or mean payoff $x{\cdot}Ax$ in time is often referred to as the *Fundamental Theorem of Natural Selection.* This statement about the replicator dynamics generalizes to the other dynamics considered here.

The generalization is based on the concept, defined by Swinkels [**58**], of a (myopic) adjustment dynamics which satisfies $\dot{x}{\cdot}Ax \geq 0$ for all $x \in \Delta$, with equality only at equilibria. If $A = A^T$ then the mean payoff $x \cdot Ax$ is increasing for every adjustment dynamics since $(x \cdot Ax)^{\cdot} = 2\dot{x} \cdot Ax \geq 0$. It is obvious that the best response dynamics (2.2) is an adjustment dynamics and it is easy to see that the other special dynamics from section 2 are as well.

As a consequence, we obtain the following result.

THEOREM 6.1. [**20, 22**] *For every partnership game $A = A^T$, the potential function $x \cdot Ax$ increases along trajectories. Hence every trajectory of every adjustment dynamics (in particular* (2.1), (2.2), (2.7), *and* (2.13)*) converges to (a connected set of) equilibria. A strict local maximizer of $x \cdot Ax$ is asymptotically stable for every adjustment dynamics.*

Generically, equilibria are isolated. Then the above result implies convergence for each trajectory. Still continua of equilibria occur in many interesting applications, see e.g. [**45**]. Even in this case, it is known that every trajectory of the replicator dynamics converges to a rest point, and hence each interior trajectory converges to a Nash equilibrium, see e.g. [**26**, ch. 23.4] or [**27**, ch. 19.2]. It is an open problem whether the same holds for the other basic dynamics.

For the perturbed dynamics (2.5) (for a concave function v on int Δ) and (2.15) there is an analog of Theorem 6.1: A suitably perturbed potential function serves as a Lyapunov function.

THEOREM 6.2. [**22, 26**] *For every partnership game $A = A^T$:*
the function $P(x) = \frac{1}{2}x{\cdot}Ax + \varepsilon v(x)$ increases monotonically along solutions of (2.5),
the function $P(x) = \frac{1}{2}x\cdot Ax + \sum_i \varepsilon_i \log x_i$ is a Lyapunov function for (2.15).
Hence every solution converges to a connected set of rest points.

For bimatrix games the adjustment property is defined as

$$\dot{x}\cdot Ay \geq 0, \quad x\cdot B\dot{y} \geq 0.$$

A bimatrix game is a partnership/potential game if $A = B$, i.e., if both players obtain the same payoff [**26**, ch. 27.2]. Then the potential $x{\cdot}Ay$ increases monotonically along every solution of every adjustment dynamics.

For the general situation of potential games between N populations with nonlinear payoff functions see [**48**].

7. A universal Shapley example

The simplest example of persistent cycling in a game dynamics is probably the RSP game (1.4) with $b > a$ for the BR dynamics (2.2) which leads to a triangular shaped limit cycle, see figure 1 (right). Historically, Shapley [**52**] gave the first such example in the context of 3×3 bimatrix games (but it is less easy to visualize because of the 4d state space). Our six basic dynamics show a similar cycling behavior for positive definite RSP games.

But given the huge pool of adjustment dynamics, we now ask: Is there an evolutionary dynamics, which converges for each game from each initial condition to an equilibrium?

Such a dynamics is assumed to be given by a differential equation

$$\dot{x} = f(x, a(x)) \tag{7.1}$$

such that f depends continuously on the population state x and the payoff function a.

For N player binary games (each player chooses between two strategies only) general evolutionary dynamics are easy to describe:
The better of the two strategies increases, the other one decreases, i.e.,

$$\dot{x}_{i1} = -\dot{x}_{i2} > 0 \Leftrightarrow a^i(1, x^{-i}) > a^i(2, x^{-i}) \tag{7.2}$$

holds for all i at all (interior) states. Here x_{ij} denotes the frequency of strategy j used by player i, and $a^i(j, x^{-i})$ his payoff. In a common interest game where each player has the same payoff function $P(x)$, along solutions $x(t)$, $P(x(t))$ increases monotonically:

$$\dot{P} = \sum_{i=1}^{N}\sum_{k=1}^{2} a^i(k, x^{-i})\dot{x}_{ik} = \sum_{i=1}^{N}\left[a^i(1, x^{-i}) - a^i(2, x^{-i})\right]\dot{x}_{i1} \geq 0. \tag{7.3}$$

A familiy of $2 \times 2 \times 2$ games. Following [**29**], we consider 3 players, each with 2 pure strategies. The payoffs are summarized in the usual way as follows.

-1,-1,-1	$0, 0, \varepsilon$
$0, \varepsilon, 0$	$\varepsilon, 0, 0$

$\varepsilon, 0, 0$	$0, \varepsilon, 0$
$0, 0, \varepsilon$	-1,-1,-1

The first player (left payoff) chooses the row, the second chooses the column, the third (right payoff) chooses one of the matrices. For $\varepsilon \neq 0$, this game has a unique equilibrium $E = \left(\frac{1}{2}, \frac{1}{2}, \frac{1}{2}\right)$ at the centroid of the state space, the cube $[0, 1]^3$. This equilibrium is regular for all ε. For $\varepsilon > 0$, this game has a best response cycle among the six pure strategy combinations $122 \to 121 \to 221 \to 211 \to 212 \to 112 \to 122$.

For $\varepsilon = 0$, this game is a potential game: Every player gets the same payoff

$$P(x) = -x_{11}x_{21}x_{31} - x_{21}x_{22}x_{32}.$$

The minimum value of P is -1 which is attained at the two pure profiles 111 and 222. At the interior equilibrium E, its value is $P(E) = -\frac{1}{4}$. P attains its maximum value 0 at the set Γ of all profiles, where two players use opposite pure strategies, whereas the remaining player may use any mixture. All points in Γ are Nash equilibria. Small perturbations in the payoffs ($\varepsilon \neq 0$) can destroy this component of equilibria.

For every natural dynamics, $P(x(t))$ increases. If $P(x(0)) > P(E) = -\frac{1}{4}$ then $P(x(t)) \to 0$ and $x(t) \to \Gamma$. Hence Γ is an attractor (an asymptotically stable invariant set) for the dynamics, for $\varepsilon = 0$.

For small $\varepsilon > 0$, there is an attractor Γ_ε near Γ whose basin contains the set $\{x : P(x) > -\frac{1}{4} + \gamma(\varepsilon)\}$, with $\gamma(\varepsilon) \to 0$, as $\varepsilon \to 0$. This follows from the fact that attractors are upper–semicontinuous against small perturbations of the dynamics (for proofs of this fact see, e.g., [**25, 2**]). But for $\varepsilon > 0$, the only equilibrium is E.

Hence we have shown

THEOREM 7.1. [**29**] *For each dynamics satisfying the assumptions* (7.2) *and continuity in payoffs, there is an open set of games and an open set of initial conditions $x(0)$ such that $x(t)$ stays away from the set of NE, for large $t > 0$.*

Similar examples can be given as 4×4 symmetric one population games, see [**27**], and 3×3 bimatrix games, see [**29**]. The proofs follow the same lines: For $\varepsilon = 0$ these are potential games, the potential maximizer is a quadrangle or a hexagon, and this component of NE disappears for $\varepsilon \neq 0$ but continues to a nearby attractor for the dynamics.

A different general nonconvergence result is due to Hart and Mas-Colell [**18**].

For specific dynamics there are many examples with cycling and even chaotic behavior: Starting with Shapley [**52**] there are [**8, 13, 16, 47, 57**] for the best response dynamics. For other examples and a more complete list of references see [**48**, ch. 9].

References

1. Aubin, J.P. and A. Cellina: *Differential Inclusions.* Springer, Berlin. 1984.
2. Benaïm M., J. Hofbauer and S. Sorin: *Perturbation of Set–valued Dynamical Systems, with Applications to Game Theory.* Preprint 2011.
3. Berger, U.: Learning in games with strategic complementarities revisited. *J. Economic Theory* **143** (2008) 292–301.
4. Berger, U. and Hofbauer, J.: Irrational behavior in the Brown–von Neumann–Nash dynamics. *Games Economic Behavior* **56** (2006), 1–6.
5. Blume, L.E.: The statistical mechanics of strategic interaction. *Games Economic Behavior* **5** (1993), 387–424.
6. Brown, G. W.: Iterative solution of games by fictitious play. In: *Activity Analysis of Production and Allocation*, pp. 374–376. Wiley. New York. 1951.
7. Brown, G. W., von Neumann, J.: Solutions of games by differential equations. *Ann. Math. Studies* **24** (1950), 73–79.
8. Cowan, S.: *Dynamical systems arising from game theory.* Dissertation (1992), Univ. California, Berkeley.

9. Cressman, R.: *Evolutionary Dynamics and Extensive Form Games.* M.I.T. Press. 2003.
10. Cressman, R.: *Extensive Form Games, Asymmetric Games and Games with Continuous Strategy Spaces.* This Volume.
11. Echenique, F and A. Edlin: Mixed equilibria are unstable in games of strategic complements. *J. Economic Theory* **118** (2004), 61–79.
12. Fudenberg, D. and Levine, D. K.: *The Theory of Learning in Games.* MIT Press. 1998.
13. Gaunersdorfer, A. and J. Hofbauer: *Fictitious play, Shapley polygons and the replicator equation. Games Economic Behavior* **11** (1995), 279–303.
14. Gilboa, I., Matsui, A.: Social stability and equilibrium. *Econometrica* **59** (1991), 859–867.
15. Gül, F., D. Pearce and E. Stacchetti: A bound on the proportion of pure strategy equilibria equilibria in generic games. *Math. Operations Research* **18** (1993) 548–552.
16. Hahn, M.: Shapley polygons in 4×4 games. *Games* **1** (2010) 189–220.
17. Harsanyi, J. C.: Oddness of the number of equilibrium points: a new proof. *Int. J. Game Theory* **2** (1973) 235–250.
18. Hart S. and A. Mas-Colell: Uncoupled dynamics do not lead to Nash equilibrium. *American Economic Review* **93** (2003) 1830–1836.
19. Hofbauer, J.: Stability for the best response dynamics. Preprint (1995).
20. Hofbauer, J.: Imitation dynamics for games. Preprint (1995).
21. Hofbauer, J.: Evolutionary dynamics for bimatrix games: A Hamiltonian system? *J. Math. Biology* **34** (1996) 675–688.
22. Hofbauer, J.: From Nash and Brown to Maynard Smith: Equilibria, dynamics and ESS. *Selection* **1** (2000), 81–88.
23. Hofbauer, J. and W. H. Sandholm: On the global convergence of stochastic fictitious play. *Econometrica* **70** (2002), 2265–2294.
24. Hofbauer, J. and W. H. Sandholm: Stable games and their dynamics. *J. Economic Theory* **144** (2009) 1665–1693.
25. Hofbauer, J. and W. H. Sandholm: Survival of Dominated Strategies under Evolutionary Dynamics. *Theor. Economics* (2011), to appear.
26. Hofbauer, J. and K. Sigmund: *The Theory of Evolution and Dynamical Systems.* Cambridge Univ. Press. 1988.
27. Hofbauer, J. and K. Sigmund: *Evolutionary Games and Population Dynamics.* Cambridge University Press. 1998.
28. Hofbauer, J. and K. Sigmund: Evolutionary Game Dynamics. *Bull. Amer. Math. Soc.* **40** (2003) 479–519.
29. Hofbauer, J. and J. Swinkels: *A universal Shapley example.* Preprint. 1995.
30. Hofbauer, J. and J. W. Weibull: Evolutionary selection against dominated strategies. *J. Economic Theory* **71** (1996), 558–573.
31. Hopkins, E.: A note on best response dynamics. *Games Economic Behavior* **29** (1999), 138–150.
32. Krishna, V.: *Learning in games with strategic complementarities*, Preprint 1992.
33. Kuhn, H. W. and S. Nasar (Eds.): *The Essential John Nash.* Princeton Univ. Press. 2002.
34. Lahkar, R. and W. H. Sandholm: The projection dynamic and the geometry of population games. *Games Economic Behavior* **64** (2008), 565–590.
35. Matsui, A.: Best response dynamics and socially stable strategies. *J. Economic Theory* **57** (1992), 343–362.
36. Maynard Smith, J.: *Evolution and the Theory of Games.* Cambridge University Press. 1982.
37. McKelvey, R. D. and T. D. Palfrey: Quantal response equilibria for normal form games. *Games Economic Behavior* **10** (1995), 6–38.
38. Monderer, D. and L. Shapley: *Potential games.* Games Economic Behavior **14** (1996), 124–143
39. Nash, J.: Equilibrium points in N–person games. *Proc. Natl. Ac. Sci.* **36** (1950), 48–49.
40. Nash, J.: *Non-cooperative games.* Dissertation, Princeton University, Dept. Mathematics. 1950. Published in [**33**].
41. Nash, J.: Non-cooperative games. *Ann. Math.* **54** (1951), 287–295.
42. Nikaido, H.: Stability of equilibrium by the Brown–von Neumann differential equation. *Econometrica* **27** (1959), 654–671.
43. Ochea, M.I.: *Essays on nonlinear evolutionary game dynamics.* Ph.D. Thesis. University of Amsterdam. 2010. http://dare.uva.nl/document/157994

44. Plank, M.: Some qualitative differences between the replicator dynamics of two player and n player games. *Nonlinear Analysis* **30** (1997), 1411–1417.
45. Pawlowitsch, C.: Why evolution does not always lead to an optimal signaling system. *Games Econ. Behav.* **63** (2008) 203–226.
46. Rahimi, M.: *Innovative Dynamics for Bimatrix Games.* Diplomarbeit. Univ. Vienna. 2009. http://othes.univie.ac.at/7816/
47. Rosenmüller, J.: Über Periodizitätseigenschaften spieltheoretischer Lernprozesse. *Z. Wahrscheinlichkeitstheorie Verw. Geb.* **17** (1971) 259–308.
48. Sandholm, W. H.: *Population Games and Evolutionary Dynamics.* MIT Press, Cambridge. 2010.
49. Sandholm, W. H.: *Stochastic Evolutionary Game Dynamics: Foundations, Deterministic Approximation, and Equilibrium Selection.* This volume.
50. Sandholm, W.H., E. Dokumaci, and F. Franchetti: *Dynamo: Diagrams for Evolutionary Game Dynamics.* 2011. http://www.ssc.wisc.edu/ whs/dynamo.
51. Schlag, K.H.: Why imitate, and if so, how? A boundedly rational approach to multi-armed bandits, *J. Economic Theory* **78** (1997) 130–156.
52. Shapley, L.: Some topics in two-person games. *Ann. Math. Studies* **5** (1964), 1–28.
53. Sigmund, K.: *Introduction to evolutionary game theory.* This volume
54. Smith, H.: *Monotone Dynamical Systems: An Introduction to the Theory of Competitive and Cooperative Systems.* Amer. Math. Soc. Math. Surveys and Monographs, Vol.41 (1995).
55. Smith, M. J.: The stability of a dynamic model of traffic assignment—an application of a method of Lyapunov. *Transportation Science* **18** (1984) 245–252.
56. Sorin, S.: *Some global and unilateral adaptive dynamics.* This volume.
57. Sparrow, C., S. van Strien and C. Harris: Fictitious play in 3×3 games: the transition between periodic and chaotic behavior. *Games and Economic Behavior* **63** (2008) 259–291.
58. Swinkels, J. M.: Adjustment dynamics and rational play in games. *Games Economic Behavior* **5** (1993), 455–484.
59. Topkis, D. M.: *Supermodularity and Complementarity.* Princeton University Press. 1998.
60. Viossat Y.: The replicator dynamics does not lead to correlated equilibria, *Games and Economic Behavior* **59** (2007), 397–407.
61. Weibull, J. W.: *Evolutionary Game Theory.* MIT Press. 1995.

Department of Mathematics, University Vienna, Austria
E-mail address: Josef.Hofbauer@univie.ac.at

Proceedings of Symposia in Applied Mathematics
Volume **69**, 2011

On Some Global and Unilateral Adaptive Dynamics

Sylvain Sorin

ABSTRACT. The purpose of this chapter is to present some adaptive dynamics arising in strategic interactive situations. We will deal with discrete time and continuous time procedures and compare their asymptotical properties. We will also consider global or unilateral frameworks and describe the wide range of applications covered by this approach. The study starts with the discrete time fictitious play procedure and its continuous time counterpart which is the best reply dynamics. Its smooth unilateral version presents interesting consistency properties. We then analyze its connection with the time average replicator dynamics. Several results rely on the theory of stochastic approximation and basic tools are briefly presented in a last section.

1. Fictitious Play and Best Reply Dynamics

Fictitious play is one of the oldest and most famous dynamical processes introduced in game theory. It has been widely studied and is a good introduction to the field of adaptive dynamics. This procedure is due to Brown (1949, 1951) and corresponds to an interactive adjustment process with (increasing and unbounded) memory.

1.1. Discrete fictitious play.

Consider a game in strategic form with a finite set of players $i \in I$, each having a finite pure strategy set S^i. For each $i \in I$, the mixed strategy set $X^i = \Delta(S^i)$ corresponds to the simplex on S^i. $F^i : S = \prod_{j \in I} S^j \to \mathbb{R}$ is the payoff of player i and we define $F^i(y) = \mathsf{E}_y F(s)$ for every $y \in \Delta(S)$, where E stands for the expectation.
The game is played repeatedly in discrete time. Given an n-stage history, which is the sequence of profiles of past moves of the players, $h_n = (x_1 = \{x_1^i\}_{i=1,\dots,I}, x_2, \dots, x_n) \in S^n$, the fictitious play procedure requires the move x_{n+1}^i of each player i at stage $n+1$ to be a best reply to the "time average moves" of her opponents.
There are two variants, that coincide in the case of two-player games :
- **independent FP**: for each i, let

$$\overline{x}_n^i = \frac{1}{n} \sum\nolimits_{m=1}^{n} x_m^i$$

1991 *Mathematics Subject Classification.* Primary 91A22.
I want to thank Bill Sandholm for useful comments.
This research was partially supported by grant ANR-08-BLAN-0294-01 (France).

and $\overline{x}_n^{-i} = \{\overline{x}_n^j\}_{j\neq i}$. Player i computes, at each stage n and for each of her opponents $j \in I$, the empirical distribution of her past moves and considers the product distribution. Then, her next move at stage $n+1$ satisfies:

$$x_{n+1}^i \in BR^i(\overline{x}_n^{-i}) \tag{1.1}$$

where BR^i denotes the best reply correspondence of player i, from $\Delta(S^{-i})$ to X^i, with $S^{-i} = \prod_{j\neq i} S^i$: $BR^i(y^{-i}) = \{x^i \in X^i; F^i(x^i, y^{-i}) = \max_{z^i\in X^i} F^i(z^i, y^{-i})\}$.
- **correlated FP**: one defines a point $\tilde{x}_n^{-i}$ in $\Delta(S^{-i})$ by :

$$\tilde{x}_n^{-i} = \frac{1}{n}\sum\nolimits_{m=1}^{n} x_m^{-i}$$

which is the empirical distribution of the joint moves of the opponents $-i$ of player i. Here the discrete time process satisfies:

$$x_{n+1}^i \in BR^i(\tilde{x}_n^{-i}). \tag{1.2}$$

Since one deals with time averages one has

$$\overline{x}_{n+1}^i = \frac{n\overline{x}_n^i + x_{n+1}^i}{n+1}$$

hence the stage difference is expressed as

$$\overline{x}_{n+1}^i - \overline{x}_n^i = \frac{x_{n+1}^i - \overline{x}_n^i}{n+1}$$

so that (1.1) can also be written as :

$$\overline{x}_{n+1}^i - \overline{x}_n^i \in \frac{1}{(n+1)}[BR^i(\overline{x}_n^{-i}) - \overline{x}_n^i]. \tag{1.3}$$

Definition. A sequence $\{x_n\}$ of moves in S satisfies **discrete fictitious play (DFP)** if (1.3) holds.

Remarks.

x_n^i does not appear explicitly any more in (1.3): the natural state variable of the process is $\overline{x}_n$ which is the product of the marginal empirical averages $\overline{x}_n^j \in X^j$.

One can define a procedure based, for each player, on her past vector payoffs $g_n^i = \{F^i(s^i, x_n^{-i})\}_{s^i\in S^i} \in \mathbb{R}^{S^i}$, rather than on the past moves of all players, as follows: $x_{n+1}^i \in \mathbf{br}^i(\bar{g}_n^i)$ with $\mathbf{br}^i(U) = \operatorname{argmax}_{X^i}\langle x, U\rangle$ and $\bar{g}_n^i = \frac{1}{n}\sum_{m=1}^n g_m^i$. Due to the linearity of the payoffs, this corresponds to the correlated fictitious play procedure. Note that $\overline{x}_n$ is no longer the common state variable but rather the correlated empirical distribution of moves $\tilde{x}_n$ which satisfies:

$$\tilde{x}_{n+1} = \frac{n\tilde{x}_n + x_{n+1}}{n+1}$$

and has the same marginal on each factor space X^i. The joint process (1.2) is defined by:

$$\tilde{x}_{n+1} - \tilde{x}_n \in \frac{1}{(n+1)}[\prod_i BR^i(\tilde{x}_n) - \tilde{x}_n]. \tag{1.4}$$

1.2. Continuous fictitious play and best reply dynamics.

The continuous time (formal) counterpart of the above difference inclusion (1.3) is the differential inclusion, called **continuous fictitious play (CFP)**:

$$\dot{X}_t^i \in \frac{1}{t}[BR^i(X_t^{-i}) - X_t^i]. \tag{1.5}$$

The change of time $Z_s = X_{e^s}$ leads to

$$\dot{Z}_s^i \in [BR^i(Z_s^{-i}) - Z_s^i] \tag{1.6}$$

which is the **(continuous time) best reply dynamics (CBR)** introduced by Gilboa and Matsui (1991), see Section 12 in K. Sigmund's chapter.
Note that the asymptotic properties of (CFP) or (CBR) are the same, since the differential inclusions differ only by their time scales.
The interpretation of (CBR) in evolutionary game theory is as follows: at each stage n a randomly selected fraction ε of the current population Z_n dies and is replaced by newborns Y_{n+1} selected according to their abilities to adjust to the current population. The discrete time process is thus

$$Z_{n+1} = \varepsilon Y_{n+1} + (1-\varepsilon)Z_n$$

with $Y_{n+1} \in BR(Z_n)$ leading to the difference inclusion

$$Z_{n+1} - Z_n \in \varepsilon[BR(Z_n) - Z_n].$$

Note that it is delicate in his framework to justify the fact that the step size ε (which is induced by the choice of the time unit) should go to 0. However numerous asymptotic results are available for small step sizes.

Comments. Recall that a solution of a differential inclusion of the form

$$\dot{z}_t \in \Psi(z_t) \tag{1.7}$$

where Ψ is a correspondence defined on a subset of $\mathbb{R}^n$ with values in $\mathbb{R}^n$, is an absolutely continuous function z from $\mathbb{R}$ to $\mathbb{R}^n$ that satisfies (1.7) almost everywhere. Let Z be a compact convex subset of $\mathbb{R}^n$ and $\Phi : Z \rightrightarrows Z$ a correspondence from Z to itself, upper semi continuous and with non empty convex values. Consider the differential inclusion

$$\dot{z}_t \in \Phi(z_t) - z_t. \tag{1.8}$$

LEMMA 1.1. *For every $z(0) \in Z$, (1.8) has a solution with $z_t \in Z$ and $z_0 = z(0)$.*

See e.g. Aubin and Cellina (1984).
In particular this applies to (CBR) where $Z = \prod X^i$ is the product of the sets of mixed strategies.
Note also that rest points of (1.8) coincide with fixed points of Φ.

1.3. General properties.

We recall briefly here basic properties of (DFP) or (CFP), in particular the link to Nash equilibrium.
Definition. A process z_n (discrete) or z_t (continuous) converges to a subset Z of some metric space if $d(z_n, Z)$ or $d(z_t, Z)$ goes to 0 as n or $t \to \infty$.

PROPOSITION 1.1. *If (DFP) or (CFP) converges to a point x, x is a Nash equilibrium.*

PROOF. If x is not a Nash equilibrium then $d(x, BR(x)) = \delta > 0$. Hence by upper semicontinuity of the best reply correspondence $d(y, BR(z)) \geq \delta/2 > 0$ for each y and z in a neighborhood of x which prevents convergence of the discrete time or continuous time processes. □

The dual property is clear:

PROPOSITION 1.2. *If x is a Nash equilibrium, it is a rest point of (CFP).*

Comments.
(i) (DFP) is "predictable": in the game with payoffs

$\sqrt{2}$	0
0	1

if player 1 follows (DFP) her move is always pure, since the past frequency of Left, say y, is a rational number so that $y\sqrt{2} = 1 - y$ is impossible; hence player 1 is guaranteed only 0. It follows that the unilateral (DFP) process has bad properties, see Section 2.
(ii) Note also the difference between convergence of the marginal distribution and convergence of the product distribution of the moves and in particular the consequences in terms of payoffs. In the next game

	L	R
T	1	0
B	0	1

a sequence of $TR, BL, TR, ...$ induces asymptotical average marginal distributions $(1/2, 1/2)$ for both players (hence optimal strategies) but the average payoff is 0 while an alternative sequence $TL, BR, ...$ would have the same average marginal distributions and payoff 1.

We analyze now (DFP) and (CFP) in some classes of games. We will deduce properties of the initial discrete time process from the analysis of the continuous time counterpart.

1.4 Zero-sum games.

This is the framework in which (DFP) was initialy introduced in order to generate optimal strategies. The continuous time model is mathematically easier to analyze.

1.4.1. *Continuous time.*
We first consider the finite case.
1) *Finite case* : Harris (1998); Hofbauer (1995); Hofbauer and Sandholm (2009). The game is defined by a bilinear map $F = F^1 = -F^2$ on a product of simplexes $X \times Y$.
Introduce $a(y) = \max_{x \in X} F(x, y)$ and $b(x) = \min_{y \in Y} F(x, y)$ that correspond to the best reply levels, then the duality gap at (x, y) is $W(x, y) = a(y) - b(x) \geq 0$. Moreover (x^*, y^*) belongs to the set of optimal strategies, $X_F \times Y_F$, iff $W(x^*, y^*) = 0$, see Section 3 of K. Sigmund's chapter. Consider the evaluation of the duality gap $W(x_t, y_t)$ along a trajectory of (1.5).

PROPOSITION 1.3. *The "duality gap" criteria converges to 0 at a speed of* $1/t$ *in (CFP).*

PROOF. Let (x_t, y_t) be a solution of (CBR) (1.6) and introduce

$$\alpha_t = x_t + \dot{x}_t \in BR^1(y_t)$$
$$\beta_t = y_t + \dot{y}_t \in BR^2(x_t).$$

The duality gap along the trajectory is given by $w_t = W(x_t, y_t)$. Note that $a(y_t) = F(\alpha_t, y_t)$ hence taking derivative with respect to the time

$$\frac{d}{dt}a(y_t) = D_1F(\alpha_t, y_t)\dot{\alpha}_t + D_2F(\alpha_t, y_t)\dot{y}_t$$

but the first term is 0 (envelope theorem). As for the second one

$$D_2F(\alpha_t, y_t)\dot{y}_t = F(\alpha_t, \dot{y}_t)$$

by linearity. Thus:

$$\dot{w}_t = F(\alpha_t, \dot{y}_t) - F(\dot{x}_t, \beta_t) = F(x_t, \dot{y}_t) - F(\dot{x}_t, y_t)$$
$$= F(x_t, \beta_t) - F(\alpha_t, y_t) = b(x_t) - a(y_t) = -w_t.$$

It follows that exponential convergence holds for (CBR)

$$w_t = e^{-t}w_0$$

hence convergence at a rate $1/t$ in the original (CFP). □

This proof in particular implies the minmax theorem and is reminiscent of the analysis due to Brown and von Neumann (1950).

The analysis extends to the framework of continuous strategy space as follows.
2) *Saddle case* : Hofbauer and Sorin (2006)
Define the condition (H) : F is a continuous, concave/convex real function defined on a product $X \times Y$ of two compact convex subsets of an euclidean space.

PROPOSITION 1.4. *Under (H), any solution w_t of (CBR) satisfies*

$$\dot{w}_t \leq -w_t \ a.e.$$

The proof, while much more involved, is in the spirit of Proposition 1.3 and the main application is (see Section 5 for the definitions):

COROLLARY 1.1. *For (CBR)*
i) $X_F \times Y_F$ is a global attractor .
ii) $X_F \times Y_F$ is a maximal invariant subset.

PROOF. From the previous Proposition 1.4 one deduces the following property: $\forall \varepsilon > 0$, $\exists T$ such that for all (x_0, y_0), $t \geq T$ implies

$$w_t \leq \varepsilon$$

hence in particular the value v_F of the game F exists and for $t \geq T$

$$b(x_t) \geq v_F - \varepsilon.$$

Continuity of F (and hence of the function b) and compactness of X imply that for any $\delta > 0$, there exists T' such that $d(x_t, X_F) \leq \delta$ as soon as $t \geq T'$. This shows that $X_F \times Y_F$ is a global attractor.
Now consider any invariant trajectory. By Proposition 1.4 at each point w one can write, for any t, $w = w_t \leq e^{-t}w_0$, but the duality gap w_0 is bounded, hence w equal to 0 which gives ii). □

To deduce properties of the discrete time process we introduce a general procedure.

1.4.2. *Discrete deterministic approximation.*
Consider again the framework of (1.8).
Let α_n a sequence of positive real numbers with $\sum \alpha_n = +\infty$.
Given $a_0 \in Z$, define inductively a_n through the following difference inclusion:

$$a_{n+1} - a_n \in \alpha_{n+1}[\Phi(a_n) - a_n]. \tag{1.9}$$

The interpretation is that the evolution of the process satisfies $a_{n+1} = \alpha_{n+1}\tilde{a}_{n+1} + (1-\alpha_{n+1})a_n$ with some $\tilde{a}_{n+1} \in \Phi(a_n)$, and where α_{n+1} is the step size at stage $n+1$.

Definition. A sequence $\{a_n\} \in Z$ following (1.9) is a **discrete deterministic approximation** (DDA) of (1.8).
The associated continuous time trajectory $\mathbf{A} : \mathbb{R}^+ \to Z$ is constructed in two stages. First define inductively a sequence of times $\{\tau_n\}$ by: $\tau_0 = 0$, $\tau_{n+1} = \tau_n + \alpha_{n+1}$; then let $A_{\tau_n} = a_n$ and extend the trajectory by linear interpolation on each interval $[\tau_n, \tau_{n+1}]$:

$$A_t = a_n + \frac{(t-\tau_n)}{(\tau_{n+1}-\tau_n)}(a_{n+1} - a_n).$$

Since $\sum \alpha_n = +\infty$ the trajectory is defined on $\mathbb{R}^+$.

To compare $\mathbf{A}$ to a solution of (1.8) we will need the approximation property corresponding to the next proposition: it states that two differential inclusions defined by correspondences having graphs close one to the other will also have sets of solutions close one to each other, on a given compact time interval.

Notations. Let $\mathcal{A}(\Phi, T, z) = \{\mathbf{z}; \mathbf{z}$ is a solution of (1.8) on $[0,T]$ with $z_0 = z\}$, $D_T(\mathbf{y}, \mathbf{z}) = \sup_{0\le t\le T} \|y_t - z_t\|$. G_Φ is the graph of Φ and G_Φ^ε is an ε-neighborhood of G_Φ.

PROPOSITION 1.5. $\forall T \ge 0, \forall \varepsilon > 0, \exists \delta > 0$ *such that*

$$\inf\{D_T(\mathbf{y}, \mathbf{z}); \mathbf{z} \in \mathcal{A}(\Phi, T, z)\} \le \varepsilon$$

for any solution $\mathbf{y}$ *of*

$$\dot{y}_t \in \tilde{\Phi}(y_t) - y_t$$

with $y_0 = z$ *and* $d(G_\Phi, G_{\tilde{\Phi}}) \le \delta$.

See e.g. Aubin and Cellina (1984), Chapter 2.

Let us now compare the two dynamics defined by $\{a_n\}$ and $\mathbf{A}$.

Case 1 Assume α_n decreasing to 0.
In this case the set $L(\{a_n\})$ of accumulation points of the sequence $\{a_n\}$ coincides with the limit set of the trajectory: $L(\mathbf{A}) = \cap_{t\ge 0}\overline{A_{[t,+\infty)}}$.

PROPOSITION 1.6.
i) If Z_0 is a global attractor for (1.8), it is also a global attractor for (1.9).
ii) If Z_0 is a maximal invariant subset for (1.8), then $L(\{a_n\}) \subset Z_0$.

PROOF. i) Given $\varepsilon > 0$, let T_1 be such that any trajectory $\mathbf{z}$ of (1.8) is within ε of Z_0 after time T_1. Given T_1 and ε, let $\delta > 0$ be defined by Proposition 1.5. Since α_n decreases to 0, given $\delta > 0$, for $n \ge N$ large enough for a_n, hence $t \ge T_2$ large enough for A_t, one can write :

$$\dot{A}_t \in \Psi(A_t) \quad \text{with} \quad G_\Psi \subset G^\delta_{\Phi - Id}.$$

Consider now A_t for some $t \geq T_1 + T_2$. Starting from any position A_{t-T_1} the continuous time process $\mathbf{z}$ defined by (1.8) approaches within ε of Z_0 at time t. Since $t - T_1 \geq T_2$, the interpolated process A_s remains within ε of the former z_s on the interval $[t - T_1, t]$, hence is within 2ε of Z_0 at time t. In particular this shows: $\forall \varepsilon, \exists N_0$ such that $n \geq N_0$ implies

$$d(a_n, Z_0) \leq 2\varepsilon.$$

ii) The result follows from the fact that $L(\mathbf{A})$ is invariant.
In fact consider $a \in L(\mathbf{A})$, hence let $t_n \to +\infty$ and $A_{t_n} \to a$. Given $T > 0$ let $\mathbf{B}^n$ denote the translated solution A_{t-t_n} defined on $[t_n - T, t_n + T]$. The sequence $\{\mathbf{B}^n\}$ of trajectories is equicontinuous and has an accumulation point $\mathbf{B}$ satisfying $B_0 = a$ and B_t is a solution of (1.8) on $[-T, +T]$. This being true for any T the result follows. □

Case 2 α_n small not vanishing.

PROPOSITION 1.7. *If Z_0 is a global attractor for (1.8), then for any $\varepsilon > 0$ there exists α such that if $\limsup_{n\to\infty} \alpha_n \leq \alpha$, there exists N with $d(a_n, Z_0) \leq \varepsilon$ for $n \geq N$. Hence a neighborhood of Z_0 is still a global attractor for (1.9).*

PROOF. The proof of Proposition 1.6 implies easily the result. □

We are now in position to study the initial discrete time fictitious play procedure.

1.4.3. *Discrete time.*
Recall that $X_F \times Y_F$ denote the product of the sets of optimal strategies in the zero-sum game with payoff F.

PROPOSITION 1.8. *(DFP) converges to $X_F \times Y_F$ in the continuous saddle zero-sum case.*

PROOF. The result follows from 1) the properties of the continuous time process, Corollary 1.1, 2) the approximation result, Proposition 1.6 and 3) the fact that the discrete time process (DFP) is a DDA of the continuous time one (CFP). □

The initial convergence result in the finite case is due to Robinson (1951). Her proof is quite involved and explicitly uses the finiteness of the strategy sets.
In this framework one has also the next result on the payoffs which is not implied by the convergence of the marginal empirical plays. In fact the distribution of the moves at each stage need not converge.

PROPOSITION 1.9. *(Rivière, 1997)*
The average of the realized payoffs along (DFP) converges to the value in the finite zero-sum case.

PROOF. Write $X = \Delta(I), Y = \Delta(J)$ and let $U_n = \sum_{p=1}^{n} F(., j_p)$ be the sum of the columns played by player 2. Consider the sum of the realized payoffs

$$R_n = \sum_{p=1}^{n} F(i_p, j_p) = \sum_{p=1}^{n} (U_p^{i_p} - U_{p-1}^{i_p})$$

Thus

$$R_n = \sum_{p=1}^{n} U_p^{i_p} - \sum_{p=1}^{n-1} U_p^{i_{p+1}} = U_n^{i_n} + \sum_{p=1}^{n-1} (U_p^{i_p} - U_p^{i_{p+1}})$$

but the fictitious property implies, since i_{p+1} is a best reply to $\bar{U}_p$, that

$$U_p^{i_p} - U_p^{i_{p+1}} \leq 0.$$

Thus $\limsup \frac{R_n}{n} \leq \limsup \max_i \frac{U_n^i}{n} \leq v$ by the previous Proposition 1.8 and the dual property implies the result. □

To summarize, in zero sum games the average empirical marginal distribution of moves are close to optimal strategies and the average payoff close to the value when the number of repetitions is large enough and both players follow (DFP).

We turn now to general I player games.

1.5. Potential games.

For a general presentation of this class, see the chapter by W. Sandholm. Since we are dealing with best-reply based processes, we can assume that the players share the same payoff function.

Hence the game is defined by a continuous payoff function F from X to $\mathbb{R}$ where each $X^i, i \in I$ is a compact convex subset of an euclidean space. Let $NE(F)$ be the set of Nash equilibria of the game defined by F.

1.5.1. *Discrete time.*

We study here the finite case and we follow Monderer and Shapley (1996).

Recall that x_n converges to $NE(F)$ if $d(x_n, NE(F))$ goes to 0. Since F is continuous and X is compact, an equivalent property is to require that for any $\varepsilon > 0$, for any n large enough x_n is an ε-equilibrium in the sense that:

$$F(x_n) + \varepsilon \geq F(x^i, x_n^{-i})$$

for all $x^i \in X^i$ and all $i \in I$.

PROPOSITION 1.10. *(DFP) converges to $NE(F)$.*

PROOF. Since F is multilinear and bounded, one has:

$$F(\bar{x}_{n+1}) - F(\bar{x}_n) = F(\bar{x}_n + \frac{1}{n+1}(x_{n+1} - \bar{x}_n)) - F(\bar{x}_n)$$

hence, by a Taylor approximation

$$F(\bar{x}_{n+1}) - F(\bar{x}_n) \geq \sum_i \frac{1}{n+1}[F(x_{n+1}^i, \bar{x}_n^{-i}) - F(\bar{x}_n)] - \frac{K_1}{(n+1)^2}$$

for some constant K_1 independent of n. Let $a_{n+1} = \sum_i [F(x_{n+1}^i, \bar{x}_n^{-i}) - F(\bar{x}_n)]$, which is ≥ 0 by definition of (DFP). Adding the previous inequality implies

$$F(\bar{x}_{n+1}) \geq \sum_{m=1}^{n+1} \frac{a_m}{m} - K_2$$

for some constant K_2. Since $a_m \geq 0$ and F is bounded, $\sum_{m=1}^{n+1} \frac{a_m}{m}$ converges. This property in turn implies

$$\lim_{N \to \infty} \frac{1}{N} \sum_{n \leq N} a_n = 0, \tag{1.10}$$

Now a consequence of (1.10) is that, for any $\varepsilon > 0$,

$$\frac{\#\{n \leq N; \bar{x}_n \notin NE^{\varepsilon}(F)\}}{N} \to 0, \qquad \text{as } N \to \infty. \tag{1.11}$$

In fact, there exists $\delta > 0$ such that $\bar{x}_n \notin NE^{\varepsilon}(F)$ forces $a_{n+1} \geq \delta$. Inequality (1.11) in turns implies that $\bar{x}_n$ belongs to $NE^{2\varepsilon}(F)$ for n large enough. Otherwise $\bar{x}_m \notin NE^{\varepsilon}(F)$ for all m in a neighborhood of n of non negligible relative size of the order $O(\varepsilon)$. (This is a general property of Cesaro mean of Cesaro means). □

1.5.2. *Continuous time.*
The finite case was studied in Harris (1998), the compact case in Benaim, Hofbauer and Sorin (2005).
Let (H') be the following hypothesis: F is defined on a product X of compact convex subsets X^i of a euclidean space, $\mathcal{C}^1$ and concave in each variable.

PROPOSITION 1.11. *Under (H'), (CBR) converges to NE(F).*

PROOF. Let $W(x) = \sum_i [G^i(x) - F(x)]$ where $G^i(x) = \max_{s \in X^i} F(s, x^{-i})$. Thus x is a Nash equilibrium iff $W(x) = 0$. Let x_t be a solution of (CBR) and consider $f_t = F(x_t)$. Then $\dot{f}_t = \sum_i D_i F(x_t)\dot{x}_t^i$. By concavity one obtains:

$$F(x_t^i, x_t^{-i}) + D_i F(x_t^i, x_t^{-i})\dot{x}_t^i \geq F(x_t^i + \dot{x}_t^i, x_t^{-i})$$

which implies

$$\dot{f}_t \geq \sum_i [F(x_t^i + \dot{x}_t^i, x_t^{-i}) - F(x_t)] = W(x_t) \geq 0$$

hence f is increasing but bounded. f is thus constant on the limit set $L(\mathbf{x})$. By the previous inequality, for any accumulation point x^* one has $W(x^*) = 0$ and x^* is a Nash equilibrium. □

In this framework also, one can deduce the convergence of the discrete time process from the properties of the continuous time analog, however $NE(F)$ is not a global attractor and the proof is much more involved (Benaim, Hofbauer and Sorin, 2005).

PROPOSITION 1.12. *Assume $F(X_F)$ with non empty interior. Then (DFP) converges to $NE(F)$.*

PROOF. Contrary to the zero-sum case where $X_F \times Y_F$ was a global attractor the proof uses here the tools of stochastic approximation, see Section 5, Proposition 5.3, with $-F$ as Lyapounov function and $NE(F)$ as critical set and Theorem 5.3. □

Remarks. Note that one cannot expect uniform convergence. See the standard symmetric coordination game:

$(1,1)$	$(0,0)$
$(0,0)$	$(1,1)$

The only attractor that contains $NE(F)$ is the diagonal. In particular convergence of (CFP) does not imply directly convergence of (DFP). Note that the equilibrium $(1/2, 1/2)$ is unstable but the time to go from $(1/2^+, 1/2^-)$ to $(1,0)$ is not bounded.

1.6. Complements.

We assume here the payoff to be multilinear and we state several properties of (DFP) and (CFP).

1.6.1. *General properties.*
Strict Nash are asymptotically stable and stricly dominated strategies are eliminated.

1.6.2. *Anticipated and realized payoff.*
Monderer, Samet and Sela (1997) introduce a comparison between the anticipated payoff at stage n ($E^i_n = F^i(x^i_n, \bar{x}^{-i}_{n-1})$) and the average payoff up to stage n (exclusive) ($A^i_n = \frac{1}{n-1}\sum_{p=1}^{n-1} F^i(x_p)$).

PROPOSITION 1.13. *Assume (DFP) for player i (with 2 players or correlated (DFP)), then*

$$E^i_n \geq A^i_n. \tag{1.12}$$

PROOF. In fact, by definition of (DFP) and by linearity:

$$\sum_{m\leq n-1} F^i(x^i_n, x^{-i}_m) \geq \sum_{m\leq n-1} F^i(s, x^{-i}_m), \quad \forall s \in X^i. \tag{1.13}$$

Write $(n-1)E^i_n = b_n = \sum_{m\leq n-1} a(n,m)$ for the left hand side. By choosing $s = x^i_{n-1}$ one obtains

$$b_n \geq a(n-1, n-1) + b_{n-1}$$

hence by induction

$$E^i_n \geq A^i_n = \sum_{m\leq n-1} a(m,m)/(n-1).$$

□

Remark
This is a unilateral property: no hypothesis is made on the behavior of player $-i$.

COROLLARY 1.2. *The average payoffs converge to the value for (DFP) in the zero-sum case.*

PROOF. Recall that in this case E^1_n (resp. E^2_n) converges to v (resp. $-v$), since $\bar{x}^{-i}_n$ converges to the set of optimal strategies of $-i$. □

The corresponding result in the continuous time setting is

PROPOSITION 1.14. *Assume (CFP) for player i in a two-person game, then*

$$\lim_{t\to+\infty} (E^i_t - A^i_t) = 0.$$

PROOF. Denote by α_s the move at time s so that:

$$tx_t = \int_0^t \alpha_s ds.$$

and $\alpha_t \in BR^1(y_t)$. One has

$$t\dot{x}_t + x_t = \alpha_t$$

which is

$$\dot{x}_t \in \frac{1}{t}[BR^1(y_t) - x_t].$$

Hence the anticipated payoff for player 1 is

$$E^1_t = F^1(\alpha_t, y_t)$$

and the past average payoff satisfies

$$tA_t^1 = \int_0^t F^1(\alpha_s, \beta_s)ds.$$

Taking derivatives one obtains

$$\frac{d}{dt}[tA_t^1] = F^1(x_t + t\dot{x}_t, y_t + t\dot{y}_t) = F^1(\alpha_t, \beta_t)$$

$$\frac{d}{dt}[tE_t^1] = E_t^1 + t\frac{d}{dt}E_t^1.$$

But $D_1F^1(\alpha, y) = 0$ (envelope theorem) and $D_2F^1(\alpha, y)\dot{y} = F^1(\alpha, \dot{y})$ by linearity. Using again linearity one obtains

$$\frac{d}{dt}[tE_t^1] = F^1(x_t + t\dot{x}_t, y_t) + F^1(x_t + t\dot{x}_t, t\dot{y}_t) = \frac{d}{dt}[tA_t^1]$$

hence there exists C such that

$$E_t - A_t = \frac{C}{t}.$$

□

COROLLARY 1.3. *Convergence of the average payoffs to the value holds for (CFP) in the zero-sum case.*

PROOF. Since y_t converges to Y_F, E_t^1 and the average payoff converges to the value. □

1.6.3. *Improvement principle.*
An interesting property is due to Monderer and Sela (1993). Note that it is not expressed in the usual state variable ($\overline{x}_n$) but is related to **Myopic Adjustment Dynamics** satisfying: $F(\dot{x}, x) \geq 0$.

PROPOSITION 1.15. *Assume (DFP) for player i with 2 players; then*

$$F^i(x_n^i, x_{n-1}^{-i}) \geq F^i(x_{n-1}). \tag{1.14}$$

PROOF. In fact the (DFP) property implies

$$F^i(x_{n-1}^i, \overline{x}_{n-2}^{-i}) \geq F^i(x_n^i, \overline{x}_{n-2}^{-i}) \tag{1.15}$$

and

$$F^i(x_n^i, \overline{x}_{n-1}^{-i}) \geq F^i(x_{n-1}^i, \overline{x}_{n-1}^{-i}). \tag{1.16}$$

Hence if equation (1.14) is not satisfied adding it to (1.15) and using the linearity of the payoff would contradict (1.16). □

These properties will be useful in proving non convergence.

1.7. Shapley's example.

Consider the next two player game, due to Shapley (1964):

$$
\mathrm{G} = \begin{array}{|c|c|c|}\hline (0,0) & (a,b) & (b,a) \\ \hline (b,a) & (0,0) & (a,b) \\ \hline (a,b) & (b,a) & (0,0) \\ \hline \end{array}
$$

with $a > b > 0$. Note that the only equilibrium is $(1/3, 1/3, 1/3)$.

PROPOSITION 1.16. *(DFP) does not always converge.*

PROOF.

Proof 1. Starting from a Pareto entry the improvement principle (1.14) implies that (DFP) will stay on Pareto entries. Hence the sum of the stage payoffs will always be $(a+b)$. If (DFP) converges then it converges to $(1/3, 1/3, 1/3)$ so that the anticipated payoff converges to the Nash payoff $\frac{a+b}{3}$ which contradicts inequality (1.12).

Proof 2. Add a line to the Shapley matrix G defining a new matrix

$$
\mathrm{G'} = \begin{array}{|c|c|c|}\hline (0,0) & (a,b) & (b,a) \\ \hline (b,a) & (0,0) & (a,b) \\ \hline (a,b) & (b,a) & (0,0) \\ \hline (c,0) & (c,0) & (c,0) \\ \hline \end{array}
$$

with $2c > b > c > \frac{a+b}{3}$.

By the improvement principle (1.14), starting from a Pareto entry one will stay on the Pareto set, hence line 4 will not be played so that (DFP) in G' is also (DFP) in G. If there were convergence it would be to a Nash equilibrium hence to $(1/3, 1/3, 1/3)$ in G, thus to $[(1/3, 1/3, 1/3, 0); (1/3, 1/3, 1/3)]$ in G'. But a best reply for player 1 to $(1/3, 1/3, 1/3)$ in G' is the fourth line, contradiction.

Proof 3. Following Shapley (1964) let us study explicitly the (DFP) trajectory. Starting from (12), there is a cycle : 12, 13, 23, 21, 31, 32, 12,... Let $r(ij)$ be the duration of the corresponding entry and α the vector of cumulative payoffs of player 1 at the beginning of the cycle i.e. if it occurs at stage $n+1$, given by:

$$
\alpha_i = \sum_{m=1}^{n} A_{i_m j_m}
$$

which is proportional to the payoff of move i against the empirical average $\bar{y}_n$. Thus, after $r(12)$ stages of (12) and $r(13)$ stages of (13) the new vector α' satisfies

$$
\alpha'_1 = \alpha_1 + r(12)a + r(13)b
$$
$$
\alpha'_2 = \alpha_2 + r(12)0 + r(13)a
$$

and then player 1 switches to move 2, hence one has

$$
\alpha'_2 \geq \alpha'_1
$$

but also

$$
\alpha_1 \geq \alpha_2
$$

(because 1 was played) so that

$$
\alpha'_2 - \alpha_2 \geq \alpha'_1 - \alpha_1
$$

which gives

$$r(13)(a-b) \geq r(12)a$$

and by induction at the next round

$$r'(11) \geq [\frac{a}{(a-b)}]^6 r(11)$$

so that exponential growth occurs and the empirical distribution does not converge (compare with the Shapley triangle, see Gaunersdorfer and Hofbauer (1995) and the chapter by J. Hofbauer). □

1.8. Other classes.

1.8.1. *Coordination games.*
A coordination game is a two person (square) game where each diagonal entry defines a pure Nash equilibrium. There are robust examples of coordination games where (DFP) fails to converge, Foster and Young (1998). Note that it is possible to have convergence of (DFP) and convergence of the payoffs to a non Nash payoff - like always mismatching. Better processes allow to select among the memory: choose s dates among the last m ones or work with finite memory adding a perturbation, see the survey in Young (2004).

1.8.2. *Dominance solvable games.*
Convergence properties are obtained in Milgrom and Roberts (1991).

1.8.3. *Supermodular games.*
In this class, convergence results are proved in Milgrom and Roberts (1990). For the case of strategic complementarity and diminishing marginal returns see Krishna and Sjöstrom (1997,1998), Berger (2008).

2. Unilateral Smooth Best Replies and Consistency

We consider here an unilateral process that will exhibit robust properties and which is deeply related to (CFP).

2.1. Consistency.

2.1.1. *Model and definitions.*
Consider a discrete time process $\{U_n\}$ of vectors in $\mathcal{U} = [0,1]^K$.
At each stage n, a player having observed the past realizations $U_1, ..., U_{n-1}$, chooses a component k_n in K. Then U_n is announced and the outcome at that stage is $\omega_n = U_n^{k_n}$.
A strategy σ in this prediction problem is specified by $\sigma(h_{n-1}) \in \Delta(K)$ (the simplex of $\mathbb{R}^K$) which is the probability distribution of k_n given the past history $h_{n-1} = (U_1, k_1, ..., U_{n-1}, k_{n-1})$.

External regret
The regret given $k \in K$ and $U \in \mathbb{R}^K$ is defined by the vector $R(k,U) \in \mathbb{R}^K$ with $R(k;U)^\ell = U^\ell - U^k, \ell \in K$.
Hence the evaluation at stage n is $R_n = R(k_n, U_n)$ i.e. $R_n^k = U_n^k - \omega_n$.
Given a sequence $\{u_m\}$, we define as usual $\bar{u}_n = \frac{1}{n}\sum_{m=1}^n u_m$. Hence the average external regret vector at stage n is $\overline{R}_n$ with

$$\overline{R}_n^k = \overline{U}_n^k - \overline{\omega}_n$$

It compares the actual (average) payoff to the payoff corresponding to a constant choice of a component, see Foster and Vohra (1999), Fudenberg and Levine (1995).

DEFINITION 2.1. A strategy σ satisfies *external consistency* (EC) if, for every process $\{U_m\}$:

$$\max_{k\in K}[\overline{R}_n^k]^+ \longrightarrow 0 \text{ a.s., as } n \to +\infty$$

or, equivalently $\quad \sum_{m=1}^n (U_m^k - \omega_m) \le o(n), \qquad \forall k \in K.$

Internal regret

The evaluation at stage n is given by a $K \times K$ matrix S_n defined by:

$$S_n^{k\ell} = \begin{cases} U_n^\ell - U_n^k & for\ k = k_n \\ 0 & otherwise. \end{cases}$$

Hence the average internal regret matrix is

$$\overline{S}_n^{k\ell} = \frac{1}{n} \sum_{m=1,k_m=k}^{n} (U_m^\ell - U_m^k).$$

This involves a comparison, for each component k, of the average payoff obtained on the dates where k was played, to the payoff that would have been induced by an alternative choice ℓ, see Foster and Vohra (1999), Fudenberg and Levine (1999). Note that we normalize by $\frac{1}{n}$ to ignore the scores of unfrequent moves.

DEFINITION 2.2. A strategy σ satisfies *internal consistency* (IC) if, for every process $\{U_m\}$ and every couple k, ℓ:

$$[\overline{S}_n^{k\ell}]^+ \longrightarrow 0 \text{ a.s., as } n \to +\infty$$

Note that no assumption is made on the process $\{U_n\}$ (like stationarity or the Markov property), moreover the player has no a priori beliefs on the law of $\{U_n\}$: we are not in a Bayesian framework and there is in general no learning, but adaptation.

2.1.2. *Application to games.*

Consider a finite game with $\#I$ players having action spaces $S^j, j \in I$. The game is repeated in discrete time and after each stage the previous profile of moves is announced. Each player i knows her payoff function $G^i : S = S^i \times S^{-i} \to \mathbb{R}$ and her observation is the vector of moves of her opponents, $s^{-i} \in S^{-i}$.

Fix i and let $K = S^i$. Player i knows in particular after stage n his stage payoff $\omega_n = G^i(k_n, s_n^{-i})$ as well as his vector payoff $U_n = G^i(., s_n^{-i}) \in \mathbb{R}^K$. The previous process describes precisely the situation that a player faces in a repeated game (with complete information and standard monitoring). She first has to choose her action, then she discovers the profile played and can evaluate her regret.

Introduce $z_n = \frac{1}{n}\sum_{m=1}^n s_m \in \Delta(S)$ with $s_m = \{s_m^j\}, j \in I$ which is the empirical distribution of profile of moves up to stage n so that by linearity

$$\bar{R}_n = \{G^i(k, z_n^{-i}) - G^i(z_n); k \in K\}.$$

Then we can express the property on the payoffs as a property on the moves.

σ satisfies EC is equivalent to : $z_n \to H^i$ a.s. with

$$H^i = \{z \in \Delta(S); G^i(k, z^{-i}) - G^i(z) \le 0, \forall k \in K\}.$$

H^i is the Hannan's set of player i, Hannan (1957).

Similarly $\bar{S}_n = S(z_n)$ with

$$S^{k,j}(z) = \sum_{\ell \in S^{-i}} [G^i(j,\ell) - G^i(k,\ell)] z(k,\ell)$$

and σ satisfies IC is equivalent to $z_n \to C^i$ a.s. with

$$C^i = \{z \in \Delta(S); S^{k,j}(z) \leq 0, \forall k, j \in K\}$$

This corresponds to the set of correlated distributions z where, for each move $k \in S^i$, k is a best reply of player i to the conditional distribution of z given k on S^{-i}. Note that $\cap_i C^i$ is the set of **correlated equilibrium distributions**, Aumann (1974).
In particular the existence of internally consistent procedures will provide an alternative proof of existence of correlated equilibrium distributions: consider any accumulation point of a trajectory generated by players using IC procedures.

2.2. Smooth fictitious play.

We described here a procedure that will satisfies IC. There are two connections with the previous section. First we will deduce properties of the random discrete time process from properties of a deterministic continuous time counterpart. Second the strategy is based on a smooth version of (DFP). Note that this procedure relies only on the previous observations of the process $\{U_n\}$ and not on the moves of the predictor, hence the regret needs not to be known, see Fudenberg and Levine (1995).

DEFINITION 2.3. A *smooth perturbation* of the payoff U is a map

$$V^\varepsilon(x, U) = \langle x, U\rangle + \varepsilon\rho(x),$$

with $0 < \varepsilon < \varepsilon_0$, such that:
(i) $\rho : X \to \mathbb{R}$ is a $\mathcal{C}^1$ function with uniform norm $\|\rho\| \leq 1$,
(ii) $\mathsf{argmax}_{x\in X} V^\varepsilon(., U)$ reduces to one point and defines a continuous map $\mathbf{br}^\varepsilon : U \to X$
called a *smooth best reply function*,
(iii) $D_1 V^\varepsilon(\mathbf{br}^\varepsilon(U), U).D\mathbf{br}^\varepsilon(U) = 0$
(for example $D_1 U^\varepsilon(., U)$ is 0 at $\mathbf{br}^\varepsilon(U)$).

A typical example is obtained via the entropy function

$$\rho(x) = -\sum_k x_k \log x_k. \tag{2.1}$$

which leads to the smooth perturbed best reply function

$$[\mathbf{br}^\varepsilon(U)]^k = \frac{\exp(U^k/\varepsilon)}{\sum_{j\in K}\exp(U^j/\varepsilon)}. \tag{2.2}$$

Let

$$W^\varepsilon(U) = \max_x V^\varepsilon(x, U) = V^\varepsilon(\mathbf{br}^\varepsilon(U), U)$$

that is close to the largest component of U and will be the evaluation criteria. A useful property is the following:

LEMMA 2.1. *(Fudenberg and Levine (1999))*

$$DW^\varepsilon(U) = \mathbf{br}^\varepsilon(U).$$

Let us first consider external consistency.

DEFINITION 2.4. A smooth fictitious play strategy σ^ε associated to the smooth best response function $\mathbf{br}^\varepsilon$ (in short a $\mathsf{SFP}(\varepsilon)$ strategy) is defined by:

$$\sigma^\varepsilon(h_n) = \mathbf{br}^\varepsilon(\overline{U}_n).$$

The corresponding discrete dynamics written in the spaces of both vectors and outcomes is

$$\overline{U}_{n+1} - \overline{U}_n = \frac{1}{n+1}[U_{n+1} - \overline{U}_n]. \tag{2.3}$$

$$\overline{\omega}_{n+1} - \overline{\omega}_n = \frac{1}{n+1}[\omega_{n+1} - \overline{\omega}_n]. \tag{2.4}$$

with

$$\mathsf{E}(\omega_{n+1}|\mathcal{F}_n) = \langle \mathbf{br}^\varepsilon(\overline{U}_n), U_{n+1}\rangle \tag{2.5}$$

which express the fact that the choice of the component of the unknown vector U_{n+1} is done according to $\sigma^\varepsilon(h_n) = \mathbf{br}^\varepsilon(\overline{U}_n)$.

We now use the properties of Section 5 to obtain, following Benaïm, Hofbauer and Sorin (2006):

LEMMA 2.2. *The process $(\overline{U}_n, \overline{\omega}_n)$ is a Discrete Stochastic Approximation of the differential inclusion with values in $\mathbb{R}^K \times \mathbb{R}$*

$$(\dot{\mathbf{u}}, \dot{\omega}) \in \{(U - \mathbf{u}, \langle \mathbf{br}^\varepsilon(\mathbf{u}), U\rangle - \omega); U \in \mathcal{U}\}. \tag{2.6}$$

The main property of the continuous dynamics is given by:

THEOREM 2.1. *The set $\{(u,\omega) \in \mathcal{U} \times \mathbb{R} : W^\varepsilon(u) - \omega \leq \varepsilon\}$ is a global attracting set for the continuous dynamics.*
In particular, for any $\eta > 0$, there exists $\bar{\varepsilon}$ such that for $\varepsilon \leq \bar{\varepsilon}$, $\limsup_{t\to\infty} W^\varepsilon(\mathbf{u}(t)) - \omega(t) \leq \eta$ (i.e. continuous $\mathsf{SFP}(\varepsilon)$ satisfies η-consistency).

PROOF. Let $q(t) = W^\varepsilon(\mathbf{u}(t)) - \omega(t)$.
Taking time derivative one obtains, using the previous two Lemmas:

$$\begin{aligned}
\dot{q}(t) &= DW^\varepsilon(\mathbf{u}(t)).\dot{\mathbf{u}}(t) - \dot{\omega}(t) \\
&= \langle \mathbf{br}^\varepsilon(\mathbf{u}(t)), \dot{\mathbf{u}}(t)\rangle - \dot{\omega}(t) \\
&= \langle \mathbf{br}^\varepsilon(\mathbf{u}(t)), U - \mathbf{u}(t)\rangle - (\langle \mathbf{br}^\varepsilon(\mathbf{u}(t)), U\rangle - \omega(t)) \\
&\leq -q(t) + \varepsilon.
\end{aligned}$$

so that $q(t) \leq \varepsilon + Me^{-t}$ for some constant M. □

In particular we deduce from Theorem 5.3 properties of the discrete time process:

THEOREM 2.2. *For any $\eta > 0$, there exists $\bar{\varepsilon}$ such that for $\varepsilon \leq \bar{\varepsilon}$, $\mathsf{SFP}(\varepsilon)$ is η-consistent.*

Let us now consider internal consistency.
Define $\bar{U}_n[k]$ as the average of U_m on the dates $1 \leq m \leq n$, where k was played. $\sigma(h_n)$ is now an invariant measure for the matrix defined by the columns

$$\{\mathbf{br}^\varepsilon(\bar{U}_n[k])\}_{k\in K}.$$

Properties similar to the above shows that σ satisfies IC, see Benaïm, Hofbauer and Sorin (2006).

For general properties of global smooth fictitious play procedures, see Hofbauer and Sandholm (2002).

Aternative consistent procedures can be found in Hart and Mas Colell (2000, 2003), see also Cesa-Bianchi and Lugosi (2006).

3. Best Reply and Average Replicator Dynamics

3.1. Presentation.

We follow here Hofbauer, Sorin and Viossat (2009).
Recall that in the framework of a symmetric 2 person game with $K \times K$ payoff matrix A played within a single population, the **replicator dynamics** is defined on the simplex Δ of $\mathbb{R}^K$ by

$$\dot{x}_t^k = x_t^k \left(e^k A x_t - x_t A x_t\right), \quad k \in K \qquad (RD) \tag{3.1}$$

where x_t^k denotes the frequency of strategy k at time t. It was introduced by Taylor and Jonker (1978) as the basic selection dynamics for the evolutionary games of Maynard Smith (1982).
In this framework the **best reply dynamics** is the differential inclusion on Δ

$$\dot{z}_t \in BR(z_t) - z_t, \qquad t \geq 0 \qquad (CBR) \tag{3.2}$$

which is the prototype of a population model of rational (but myopic) behaviour.

Despite the different interpretation and the different dynamic character there are amazing similarities in the long run behaviour of these two dynamics, that have been summarized in the following heuristic principle:
For many games, the long run behaviour ($t \to \infty$) of the time averages $X_t = \frac{1}{t}\int_0^t x_s ds$ of the trajectories x_t of the replicator equation is the same as for the BR trajectories.

We provide here a rigorous statement that largely explains this heuristic by showing that for any interior solution of (RD), for every $t \geq 0$, x_t is an approximate best reply against X_t and the approximation gets better as $t \to \infty$. This implies that X_t is an asymptotic pseudo trajectory of (CBR), see section 5, and hence the limit set of X_t has the same properties as a limit set of a true orbit of (CBR), i.e. it is invariant and internally chain transitive under (CBR).
The main tool to prove this is via the logit map which is a canonical smoothing of the best response correspondence. We show that x_t equals the logit approximation at X_t with error rate $\frac{1}{t}$.

3.2. Unilateral processes.

The model will be in the framework of an I-person game but we consider the dynamics for one player, without hypotheses on the behavior of the others. The framework is unilateral, as in the previous section, but now in continuous time. Hence, from the point of view of this player, she is facing a (measurable) vector outcome process $\mathcal{U} = \{U_t, t \geq 0\}$, with values in the cube $C = [-c, c]^K$ where K is her move set and c is some positive constant. U_t^k is the payoff at time t if k is the move at that time. The cumulative vector outcome up to stage t is $S_t = \int_0^t U_s ds$ and its time average is denoted $\bar{U}_t = \frac{1}{t} S_t$.
br denotes the (payoff based) best reply correspondence from C to Δ defined by

$$\mathbf{br}(U) = \{x \in \Delta; \langle x, U\rangle = \max_{y\in\Delta}\langle y, U\rangle\}.$$

The $\mathcal{U}$-best reply process (CBR) is defined on Δ by

$$\dot{X}_t \in [\mathbf{br}(\bar{U}_t) - X_t]. \tag{3.3}$$

The $\mathcal{U}$-replicator process (RP) is specified by the following equation on Δ:

$$\dot{x}_t^k = x_t^k[U_t^k - \langle x_t, U_t \rangle], \qquad k \in K. \tag{3.4}$$

Explicitly, in the framework of an I-player game with payoff for player 1 defined by a function G from $\prod_{i \in I} S^i$ to $\mathbb{R}$, with $X^i = \Delta(S^i)$, U is the vector payoff i.e. $U_t = G(., x_t^{-1})$.
If all the players follow a (payoff based) continuous time correlated fictitious play dynamics, each time average strategy satisfies (3.3).
If all the players follow the replicator dynamics then (3.4) is the replicator dynamics equation.

3.3. Logit rule and perturbed best reply.
Define the map L from $\mathbb{R}^K$ to Δ by

$$L^k(V) = \frac{\exp V^k}{\sum_j \exp V^j}. \tag{3.5}$$

Given $\eta > 0$, let $[\mathbf{br}]^\eta$ be the correspondence from C to Δ with graph being the η-neighborhood for the uniform norm of the graph of $\mathbf{br}$.
The L map and the $\mathbf{br}$ correspondence are related as follows:

PROPOSITION 3.1. *For any $U \in C$ and $\varepsilon > 0$*

$$L(U/\varepsilon) \in [\mathbf{br}]^{\eta(\varepsilon)}(U)$$

with $\eta(\varepsilon) \to 0$ as $\varepsilon \to 0$.

Remarks. L is also given by

$$L(V) = \text{argmax}_{x \in \Delta}\{\langle x, V \rangle - \sum_k x^k \log x^k\}.$$

Hence introducing the (payoff based) perturbed best reply $\mathbf{br}^\varepsilon$ from C to Δ defined by

$$\mathbf{br}^\varepsilon(U) = \text{argmax}_{x \in \Delta}\{\langle x, U \rangle - \varepsilon \sum_k x^k \log x^k\}$$

one has $L(U/\varepsilon) = \mathbf{br}^\varepsilon(U)$.
The map $\mathbf{br}^\varepsilon$ is the logit approximation, see (2.2).

3.4. Explicit representation of the replicator process.
The following procedure has been introduced in discrete time in the framework of on-line algorithms under the name "multiplicative weight algorithm", Littlestone and Warmuth (1994). We use here the name (CEW) (continuous exponential weight) for the process defined, given $\mathcal{U}$, by

$$x_t = L(\int_0^t U_s ds).$$

The main property of (CEW) that will be used is that it provides an explicit solution of (RP).

PROPOSITION 3.2. *(CEW) satisfies (RP).*

PROOF. Straightforward computations lead to

$$\dot{x}_t^k = x_t^k U_t^k - x_t^k \sum_j \frac{U_t^j \exp \int_0^t U_v^j dv}{\sum_j \exp \int_0^t U_v^j dv}$$

which is

$$\dot{x}_t^k = x_t^k [U_t^k - \langle x_t, U_t \rangle]$$

hence gives the previous (RP) equation (3.4). □

The link with the best reply correspondence is the following:

PROPOSITION 3.3. *(CEW) satisfies*

$$x_t \in [\mathbf{br}]^{\delta(t)}(\bar{U}_t)$$

with $\delta(t) \to 0$ *as* $t \to \infty$.

PROOF. Write

$$x_t = L(\int_0^t U_s ds) = L(t\, \bar{U}_t).$$

Then

$$x_t = L(U/\varepsilon) \in [\mathbf{br}]^{\eta(\varepsilon)}(U)$$

with $U = \bar{U}_t$ and $\varepsilon = 1/t$, by Proposition 3.1. Let then $\delta(t) = \eta(1/t)$. □

We describe here the consequences for the time average process. Define

$$X_t = \frac{1}{t} \int_0^t x_s ds.$$

PROPOSITION 3.4. *If* x_t *follows (CEW) then* X_t *satisfies*

$$\dot{X}_t \in \frac{1}{t}([\mathbf{br}]^{\delta(t)}(\bar{U}_t) - X_t). \tag{3.6}$$

with $\delta(t) \to 0$ *as* $t \to \infty$.

PROOF. One has, taking derivatives:

$$t\dot{X}_t + X_t = x_t$$

and the result follows from the properties of x_t. □

3.5. Consequences for games.

Consider a 2 person (bimatrix) game (A, B).
If the game is symmetric this gives rise to the single population replicator dynamics (RD) and best reply dynamics (BRD) as defined in section 1.
Otherwise, we consider the two population replicator dynamics

$$\begin{aligned} \dot{x}_t^k &= x_t^k \left(e^k A y_t - x_t A y_t\right), \quad k \in S^1 \\ \dot{y}_t^k &= y_t^k \left(x_t B e^k - x_t B y_t\right), \quad k \in S^2 \end{aligned} \tag{3.7}$$

and the corresponding BR dynamics as in (3).
Let M be the state space (a simplex Δ or a product of simplices $\Delta_1 \times \Delta_2$). We now use the previous results with the process $\mathcal{U}$ being defined by $U_t = Ay_t$ for player 1, hence $\bar{U}_t = AY_t$. Note that $\mathbf{br}(AY) = BR^1(Y)$.

PROPOSITION 3.5. *The limit set of every replicator time average process* X_t *starting from an initial point* $x_0 \in M$ *is a closed subset of* M *which is invariant and internally chain transitive under (CBR).*

PROOF. Equation (3.6) implies that X_t satisfies a perturbed version of (CFP) hence X_{e^t} is a perturbed solution to the differential inclusion (CBR), according to Section 5 and Theorems 5.1 and 5.2 apply. □

In particular this implies:

PROPOSITION 3.6. *Let $\mathcal{A}$ be the global attractor (i.e., the maximal invariant set) of (CBR). Then the limit set of every replicator time average process X_t is a subset of $\mathcal{A}$.*

3.6. External consistency.

The natural continuous time counterpart of the (discrete time) notion is the following: a procedure satisfies external consistency if for each process $\mathcal{U}$ taking values in $\mathbb{R}^K$, it produces a process $x_t \in \Delta$, such that for all k

$$\int_0^t [U_s^k - \langle x_s, U_s\rangle] ds \le C_t = o(t)$$

where, using a martingale argument, we have replaced the actual random payoff at time s by its conditional expectation $\langle x_s, U_s\rangle$. This property says that the (expected) average payoff induced by x_t along the play is asymptotically not less than the payoff obtained by any fixed choice $k \in K$.

PROPOSITION 3.7. *(RP) satisfies external consistency.*

PROOF. By integrating equation (3.4), one obtains, on the support of x_0:

$$\int_0^t [U_s^k - \langle x_s, U_s\rangle] ds = \int_0^t \frac{\dot{x}_s^k}{x_s^k} ds = \log(\frac{x_t^k}{x_0^k}) \le -\log x_0^k.$$

□

This result is the unilateral analog of the fact that interior rest points of (RD) are equilibria. A myopic unilateral adjustment process provides asymptotic optimal properties in terms of no regret.

Back to a game framework this implies that if player 1 follows (RP) the set of accumulation points of the empirical correlated distribution process will belong to her reduced Hannan set:

$$\bar{H}^1 = \{\theta \in \Delta(S); G^1(k, \theta^{-1}) \le G^1(\theta), \forall k \in S^1\}$$

with equality for at least one component.
The example due to Viossat (2007, 2008) of a game where the limit set for the replicator dynamics is disjoint from the unique correlated equilibrium shows that (RP) does not satisfy internal consistency.
This later property uses additional information that is not taken into account in the replicator dynamics. This topic deserves further study.

3.7. Comments.

We can now compare several processes in the spirit of (payoff based) fictitious play.
The original fictitious play process (I) is defined by

$$x_t \in \mathbf{br}(\bar{U}_t)$$

The corresponding time average satisfies (CFP).
With a smooth best reply process one has (II)

$$x_t = \mathbf{br}^{\varepsilon}(\bar{U}_t)$$

and the corresponding time average satisfies a smooth fictitious play process.
Finally the replicator process (III) satisfies

$$x_t = \mathbf{br}^{1/t}(\bar{U}_t)$$

and the time average follows a time dependent perturbation of the fictitious play process.

While in (I), the process x_t follows exactly the best reply correspondence, the induced average X_t does not have good unilateral properties.
One the other hand for (II), X_t satisfies a weak form of external consistency, with an error term $\alpha(\varepsilon)$ vanishing with ε.
In contrast, (III) satisfies exact external consistency due to a both smooth and time dependent approximation of **br**.

4. General Adaptive Dynamics

We consider here random processes corresponding to adaptive behavior in repeated interactions.
The analysis is done from the point of view of one player, having a finite set K of actions. Time is discrete and the behavior of the player depends upon a parameter $z \in Z$.
At stage n, the state is z_{n-1} and the process is defined by two functions:
a **decision map** σ from Z to $\Delta(K)$ (the simplex on K) defining the law π_n of the current action k_n as a function of the parameter:

$$\pi_n = \sigma(z_{n-1})$$

and given the observation ω_n of the player, after the play at stage n, an **updating rule** for the state variable, that depends upon the stage:

$$z_n = \Phi_n(z_{n-1}, \omega_n).$$

Remark
Note that the decision map is stationary but that the updating rule may depend upon the stage.
A typical assumption in game theory is that the player knows his payoff function $G : K \times L \to \mathbb{R}$ and that the observation ω is the vector of moves of his opponents, $\ell \in L$. In particular ω_n contains the stage payoff $g_n = G(k_n, \ell_n)$ as well as the vector payoff $U_n = G(., \ell_n) \in \mathbb{R}^K$.

Example 1: Fictious Play
The state space is usually the empirical distribution of actions of the opponents but one can as well take $\omega_n = U_n$, the vector payoff, then $z_n = \overline{U}_n$ is the average vector payoff thus satisfies:

$$z_n = \frac{(n-1)z_{n-1} + U_n}{n}$$

and

$$\sigma(z) \in BR(z) \qquad \text{or} \qquad \sigma(z) = BR^{\varepsilon}(z).$$

Example 2: Potential regret dynamics
Here

$$R_n = U_n - g_n \mathbf{1}$$

is the "regret vector" at stage n and the updating rule $z_n = \Phi_n(z_{n-1}, \omega_n)$ is simply

$$z_n = \overline{R}_n.$$

Choose P to be a "potential function" for the negative orthant $D = \mathbb{R}^K_-$ and for $z \notin D$ let $\sigma(z)$ be proportional to $\nabla P(z)$.

Example 3: Cumulative proportional reinforcement
The observation ω_n is only the stage payoff g_n (we assume all payoffs ≥ 1).
The updating rule is

$$z^k_n = z^k_{n-1} + g_n \, \mathbf{I}_{\{k_n = k\}}$$

and the decision map is $\sigma(z)$ proportional to the vector z.
There is an important literature on such reinforcement dynamics, see e.g. Beggs (2005), Börgers, Morales and Sarin (2004), Börgers and Sarin (1997), Hopkins (2002), Hopkins and Posch (2005), Laslier, Topol and Walliser (2001), Leslie and Collins (2005), Pemantle (2007), Posch (1997).

Note that these three procedures can be written as

$$z_n = \frac{(n-1)z_{n-1} + v_n}{n}$$

where v_n is a random variable depending on the action(s) ℓ of the opponent(s) and on the action k_n having distribution $\sigma(z_{n-1})$. Thus

$$z_n - z_{n-1} = \frac{1}{n}[v_n - z_{n-1}].$$

Write

$$v_n = E_{\pi_n}(v_n|z_1, ..., z_{n-1}) + [v_n - E_{\pi_n}(v_n|z_1, ..., z_{n-1})]$$

and define

$$S(z_{n-1}) = Co\{E_{\pi_n}(v_n|z_1, ..., z_{n-1}); \ell \in L\}$$

where Co stands for the convex hull. Thus

$$z_n - z_{n-1} \in \frac{1}{n}[S(z_{n-1}) - z_{n-1}].$$

The differential inclusion is

(4.1) $$\dot{z} \in S(z) - z$$

and the process z_n is a Discrete Stochastic Approximation of (4.1), see section 5.

For further results with explicit applications of this procedure see e.g. Hofbauer and Sanholm (2002), Benaïm, Hofbauer and Sorin (2006), Cominetti, Melo and Sorin (2010).

In conclusion, a large class of adaptive dynamics can be expressed in discrete time as a random difference equation with vanishing step size. Information on the asymptotic behavior can then be obtained by studing the continuous time deterministic analog obtained as above.

5. Stochastic Approximation for Differential Inclusions

We summarize here results from Benaïm, Hofbauer and Sorin (2005).

5.1. Differential inclusions.

Given a correspondence F from $\mathbb{R}^m$ to itself, consider the differential inclusion

$$\dot{\mathbf{x}} \in F(\mathbf{x}) \qquad (I)$$

It induces a set-valued dynamical system $\{\Phi_t\}_{t\in\mathbb{R}}$ defined by

$$\Phi_t(x) = \{\mathbf{x}(t) : \mathbf{x} \text{ is a solution to (I) with } \mathbf{x}(0) = x\}.$$

We also write $\mathbf{x}(t) = \phi_t(x)$.

DEFINITION 5.1.
1) x is a *rest point* if $0 \in F(x)$.
2) A set C is *strongly forward invariant* (SFI) if $\Phi_t(C) \subset C$ for all $t \geq 0$.
3) C is *invariant* if for any $x \in C$ there exists a complete solution: $\phi_t(x) \in C$ for all $t \in \mathbb{R}$.
4) C is *Lyapounov stable* if: $\forall\varepsilon > 0, \exists\delta > 0$ such that $d(y, C) \leq \delta$ implies $d(\Phi_t(y), C) \leq \varepsilon$ for all $t \geq 0$, i.e.

$$\Phi_{[0,+\infty)}(C^\delta) \subset C^\varepsilon.$$

5) C is a *sink* if there exists $\delta > 0$ such that for any $y \in C^\delta$ and any ϕ:

$$d(\phi_t(y), C) \to 0 \text{ as } t \to \infty.$$

A neighborhood U of C having this property is called a *basin of attraction* of C.
6) C is *attracting* if it is compact and the previous property is uniform. Thus there exist $\delta > 0$, $\varepsilon_0 > 0$ and a map $T : (0, \varepsilon_0) \to \mathbb{R}^+$ such that: for any $y \in C^\delta$, any solution ϕ, $\phi_t(y) \in C^\varepsilon$ for all $t \geq T(\varepsilon)$, i.e.

$$\Phi_{[T(\varepsilon),+\infty)}(C^\delta) \subset C^\varepsilon, \quad \forall\varepsilon \in (0, \varepsilon_0).$$

A neighborhood U of C having this property is called a *uniform basin of attraction* of C and we will write $(C; U)$ for the couple.
7) C is an *attractor* if it is attracting and invariant.
8) C is *forward precompact* if there exists a compact K and a time T such that $\Phi_{[T,+\infty)}(C) \subset K$.
9) The *ω-limit set* of C is defined by

$$\omega_\Phi(C) = \cap_{s\geq 0}\overline{\cup_{y\in C} \cup_{t\geq s} \Phi_t(y)} = \cap_{s\geq 0}\overline{\Phi_{[s,+\infty)}(C)} \tag{5.1}$$

where $\overline{A}$ denotes the closure of the set A.

DEFINITION 5.2.
i) Given a closed invariant set L, the induced set-valued dynamical system Φ^L is defined on L by

$$\Phi_t^L(x) = \{\mathbf{x}(t) : \mathbf{x} \text{ is a solution to (I) with } \mathbf{x}(0) = x \text{ and } \mathbf{x}(\mathbb{R}) \subset L\}.$$

Note that $L = \Phi_t^L(L)$ for all t.
ii) Let $A \subset L$ be an attractor for Φ^L. If $A \neq L$ and $A \neq \emptyset$, then A is a *proper attractor*.
An invariant set L is *attractor free* if Φ^L has no proper attractor.

5.2. Attractors.
The next notion is fondamental in the analysis.

DEFINITION 5.3.
C is *asymptotically stable* if it has the following properties
i) invariant
ii) Lyapounov stable
iii) sink.

PROPOSITION 5.1. *Assume C compact. Attractor is equivalent to asymptotically stable.*

PROPOSITION 5.2. *Let A be a compact set, U be a relatively compact neighborhood and V a function from $\overline{U}$ to $\mathbb{R}^+$. Consider the following properties*
i) U is (SFI)
ii) $V^{-1}(0) = A$
iii) V is continuous and strictly decreasing on trajectories on $U \setminus A$:

$$V(x) > V(y), \quad \forall x \in U \setminus A, \forall y \in \phi_t(x), \quad \forall t > 0$$

iv) V is upper semi continuous and strictly decreasing on trajectories on $\overline{U} \setminus A$.
a) Then under i), ii) and iii) A is Lyapounov stable and $(A; U)$ is attracting.
b) Under i), ii) and iv), $(B; U)$ is an attractor for some $B \subset A$.

DEFINITION 5.4.
A real continuous function V on U open in $\mathbb{R}^m$ is a *Lyapunov function for $A \subset U$* if : $V(y) < V(x)$ for all $x \in U \setminus A, y \in \phi_t(x), t > 0$; and $V(y) \leq V(x)$ for all $x \in A, y \in \phi_t(x)$ and $t \geq 0$.
Note that for each solution ϕ, V is constant along its *limit set*

$$L(\phi)(x) = \cap_{s \geq 0} \overline{\phi_{[s,+\infty)}(x)}.$$

PROPOSITION 5.3. *Suppose V is a Lyapunov function for A. Assume that $V(A)$ has empty interior. Let L be a non empty, compact, invariant and attractor free subset of U. Then L is contained in A and V is constant on L.*

5.3. Asymptotic pseudo-trajectories and internally chain transitive sets.

5.3.1. *Asymptotic pseudo-trajectories.*

DEFINITION 5.5. The *translation flow* $\Theta : C^0(\mathbb{R}, \mathbb{R}^m) \times \mathbb{R} \to C^0(\mathbb{R}, \mathbb{R}^m)$ is defined by

$$\Theta_t(\mathbf{x})(s) = \mathbf{x}(s+t).$$

A continuous function $\mathbf{z} : \mathbb{R}^+ \to \mathbb{R}^m$ is an *asymptotic pseudo-trajectory* (APT) for Φ if for all T

$$\lim_{t \to \infty} \inf_{\mathbf{x} \in S_{\mathbf{z}(t)}} \sup_{0 \leq s \leq T} \|\mathbf{z}(t+s) - \mathbf{x}(s)\| = 0. \tag{5.2}$$

where S_x denotes the set of all solutions of (I) starting from x at 0 and $S = \bigcup_{x \in \mathbb{R}^m} S_x$.

In other words, for each fixed T, the curve: $s \to \mathbf{z}(t+s)$ from $[0, T]$ to $\mathbb{R}^m$ shadows some trajectory for (I) of the point $\mathbf{z}(t)$ over the interval $[0, T]$ with arbitrary accuracy, for sufficiently large t. Hence $\mathbf{z}$ has a forward trajectory under Θ attracted by S. One extends $\mathbf{z}$ to $\mathbb{R}$ by letting $\mathbf{z}(t) = \mathbf{z}(0)$ for $t < 0$.

5.3.2. *Internally chain transitive sets.*
Given a set $A \subset \mathbb{R}^m$ and $x, y \in A$, we write $x \to_A y$ if for every $\varepsilon > 0$ and $T > 0$ there exists an integer $n \in I\!N$, solutions $\mathbf{x}_1, \ldots, \mathbf{x}_n$ to (I), and real numbers $t_1, t_2, \ldots, t_n$ greater than T such that
a) $\mathbf{x}_i(s) \in A$ for all $0 \leq s \leq t_i$ and for all $i = 1, \ldots, n$,
b) $\|\mathbf{x}_i(t_i) - \mathbf{x}_{i+1}(0)\| \leq \varepsilon$ for all $i = 1, \ldots, n-1$,
c) $\|\mathbf{x}_1(0) - x\| \leq \varepsilon$ and $\|\mathbf{x}_n(t_n) - y\| \leq \varepsilon$.
The sequence $(\mathbf{x}_1, \ldots, \mathbf{x}_n)$ is called an (ε, T) chain (in A from x to y) for (I).

DEFINITION 5.6.
A set $A \subset \mathbb{R}^m$ is *internally chain transitive* (ICT) if it is compact and $x \to_A y$ for all $x, y \in A$.

LEMMA 5.1. *An internally chain transitive set is invariant.*

PROPOSITION 5.4. *Let L be internally chain transitive. Then L has no proper attracting set for Φ^L.*

This (ICT) notion of recurrence due to Conley (1978) for classical dynamical systems is well suited to the description of the asymptotic behavior of APT, as shown by the following theorem. Let

$$L(\mathbf{z}) = \bigcap_{t \geq 0} \overline{\{\mathbf{z}(s) : s \geq t\}}$$

be the limit set.

THEOREM 5.1. *Let $\mathbf{z}$ be a bounded APT of (I). Then $L(\mathbf{z})$ is internally chain transitive.*

5.4. Perturbed solutions.
The purpose of this paragraph is to study trajectories which are obtained as (deterministic or random) perturbations of solutions of (I).
5.4.1. *Perturbed solutions.*

DEFINITION 5.7.
A continuous function $\mathbf{y} : \mathbb{R}_+ = [0, \infty) \to \mathbb{R}^m$ is a *perturbed solution* to (I) if it satisfies the following set of conditions (II):
i) $\mathbf{y}$ is absolutely continuous.
ii) There exists a locally integrable function $t \mapsto U(t)$ such that

$$\lim_{t \to \infty} \sup_{0 \leq v \leq T} \left\| \int_t^{t+v} U(s)\, ds \right\| = 0$$

for all $T > 0$
iii)

$$\frac{d\mathbf{y}(t)}{dt} - U(t) \in F^{\delta(t)}(\mathbf{y}(t))$$

for almost every $t > 0$, for some function $\delta : [0, \infty) \to \mathbb{R}$ with $\delta(t) \to 0$ as $t \to \infty$. Here $F^\delta(x) := \{y \in \mathbb{R}^m : \exists z : \|z - x\| < \delta,\ d(y, F(z)) < \delta\}$.

The aim is to investigate the long-term behavior of $\mathbf{y}$ and to describe its limit set $L(\mathbf{y})$ in terms of the dynamics induced by F.

THEOREM 5.2. *Any bounded solution $\mathbf{y}$ of (II) is an APT of (I).*

5.4.2. *Discrete stochastic approximation.*
As will be shown here, a natural class of perturbed solutions to F arises from certain stochastic approximation processes.

DEFINITION 5.8.
A discrete time process $\{x_n\}_{n\in \mathbb{N}}$ with values in $\mathbb{R}^m$ is a solution for (III) if it verifies a recursion of the form

$$x_{n+1} - x_n - \gamma_{n+1}U_{n+1} \in \gamma_{n+1}F(x_n), \qquad (III)$$

where the characteristics γ and U satisfy
i) $\{\gamma_n\}_{n\geq 1}$ is a sequence of nonnegative numbers such that

$$\sum_n \gamma_n = \infty, \qquad \lim_{n\to\infty} \gamma_n = 0;$$

ii) $U_n \in \mathbb{R}^m$ are (deterministic or random) perturbations.

To such a process is naturally associated a continuous time process as follows.

DEFINITION 5.9.
Let $\tau_0 = 0$ and $\tau_n = \sum_{i=1}^n \gamma_i$ for $n \geq 1$, and define the continuous time affine interpolated process $\mathbf{w} : \mathbb{R}_+ \to \mathbb{R}^m$ by

$$\mathbf{w}(\tau_n + s) = x_n + s\frac{x_{n+1} - x_n}{\tau_{n+1} - \tau_n}, \quad s \in [0, \gamma_{n+1}). \qquad (IV)$$

5.5. From interpolated process to perturbed solutions.
The next result gives sufficient conditions on the characteristics of the discrete process (III) for its interpolation (IV) to be a perturbed solution (II).
If (U_i) are random variables, assumptions (i) and (ii) below hold with probability one.

PROPOSITION 5.5. *Assume that the following hold:*
(i) *For all* $T > 0$

$$\lim_{n\to\infty} \sup \left\{ \left\| \sum_{i=n}^{k-1} \gamma_{i+1}U_{i+1} \right\| : k = n+1, \dots, m(\tau_n + T) \right\} = 0,$$

where

$$m(t) = \sup\{k \geq 0 : t \geq \tau_k\}; \tag{5.3}$$

(ii) $\sup_n \|x_n\| = M < \infty$.
Then the interpolated process $\mathbf{w}$ *is a perturbed solution of* (I).

We describe now sufficient conditions.
Let $(\Omega, \mathbf{\Psi}, P)$ be a probability space and $\{\mathbf{\Psi}_n\}_{n\geq 0}$ a filtration of $\mathbf{\Psi}$ (i.e., a nondecreasing sequence of sub-σ-algebras of $\mathbf{\Psi}$). A stochastic process $\{x_n\}$ given by (III) satisfies the *Robbins–Monro condition* with martingale difference noise if its characteristics satisfy the following:
i) $\{\gamma_n\}$ is a deterministic sequence.
ii) $\{U_n\}$ is *adapted* to $\{\mathbf{\Psi}_n\}$, which means that U_n is measurable with respect to $\mathbf{\Psi}_n$ for each $n \geq 0$.
iii) $\mathbf{E}(U_{n+1} \mid \mathbf{\Psi}_n) = 0$.
The next proposition is a classical estimate for stochastic approximation processes. Note that F does not appear, see Benaïm (1999) for a proof and further references.

PROPOSITION 5.6. *Let $\{x_n\}$ given by (III) be a Robbins–Monro process. Suppose that for some $q \geq 2$*

$$\sup_n \mathbf{E}(\|U_n\|^q) < \infty$$

and

$$\sum_n \gamma_n^{1+q/2} < \infty.$$

Then assumption (i) *of Proposition* 5.5 *holds with probability* 1.

Remark. Typical applications are
i) U_n uniformly bounded in L^2 and $\gamma_n = \frac{1}{n}$,
ii) U_n uniformly bounded and $\gamma_n = o(\frac{1}{\log n})$.

5.6. Main result.

Consider a random discrete process defined on a compact subset of $\mathbb{R}^K$ and satisfying the differential inclusion :

$$Y_n - Y_{n-1} \in a_n[T(Y_{n-1}) + W_n]$$

where
i) T is an u.s.c. correspondence with compact convex values
ii) $a_n \geq 0$, $\sum_n a_n = +\infty$, $\sum_n a_n^2 < +\infty$
iii) $E(W_n|Y_1, ..., Y_{n_1}) = 0$.

THEOREM 5.3. *The set of accumulation points of $\{Y_n\}$ is almost surely a compact set, invariant and attractor free for the dynamical system defined by the differential inclusion:*

$$\dot{Y} \in T(Y).$$

References

[1] AUBIN J.-P. AND A. CELLINA (1984) *Differential Inclusions*, Springer.

[2] AUER P., CESA-BIANCHI N., FREUND Y. AND R.E. SHAPIRE (2002) The nonstochastic multi-armed bandit problem, *SIAM J. Comput.*, **32**, 48-77.

[3] AUMANN R.J. (1974) Subjectivity and correlation in randomized strategies, *Journal of Mathematical Economics*, **1**, 67-96.

[4] BEGGS A. (2005) On the convergence of reinforcement learning, *Journal of Economic Theory*, **122**, 1-36.

[5] BENAÏM M. (1996) A dynamical system approach to stochastic approximation, *SIAM Journal on Control and Optimization*, **34**, 437-472.

[6] BENAÏM M. (1999) Dynamics of Stochastic Algorithms, *Séminaire de Probabilités*, **XXIII**, Azéma J. and alii eds, Lectures Notes in Mathematics, **1709**, Springer, 1-68.

[7] BENAÏM M. AND M.W. HIRSCH (1996) Asymptotic pseudotrajectories and chain recurrent flows, with applications, *J. Dynam. Differential Equations*, **8**, 141-176.

[8] BENAÏM M. AND M.W. HIRSCH (1999) Mixed equilibria and dynamical systems arising from fictitious play in perturbed games, *Games and Economic Behavior*, **29**, 36-72.

[9] BENAÏM M., J. HOFBAUER AND S. SORIN (2005) Stochastic approximations and differential inclusions, *SIAM J. Opt. and Control*, **44**, 328-348.

[10] BENAÏM M., J. HOFBAUER AND S. SORIN (2006) Stochastic approximations and differential inclusions. Part II: applications, *Mathematics of Operations Research*, **31**, 673-695.

[11] BERGER U. (2005) Fictitious play in $2 \times n$ games, *Journal of Economic Theory*, **120**, 139-154.

[12] BERGER U. (2008) Learning in games with strategic complementarities revisited, *Journal of Economic Theory*, **143**, 292-301.

[13] BÖRGERS T., A. MORALES AND R. SARIN (2004) Expedient and monotone learning rules, *Econometrica*, **72**, 383-406.

[14] BÖRGERS T. AND R. SARIN (1997) Learning through reinforcement and replicator dynamics, *Journal of Economic Theory*, **77**, 1-14.

[15] BROWN G. W. (1949) Some notes on computation of games solutions, RAND Report P-78, The RAND Corporation, Santa Monica, California.

[16] BROWN G. W. (1951) Iterative solution of games by fictitious play, in Koopmans T.C. (ed.) *Activity Analysis of Production and Allocation* , Wiley, 374-376.

[17] BROWN G.W. AND J. VON NEUMANN (1950) Solutions of games by differential equations, *Contributions to the Theory of Games I*, Annals of Mathematical Studies, **24**, 73-79.

[18] CESA-BIANCHI N. AND G. LUGOSI (2006) *Prediction, Learning and Games*, Cambridge University Press.

[19] COMINETTI R., E. MELO AND S. SORIN (2010) A payoff-based learning procedure and its application to traffic games, *Games and Economic Behavior*, **70**, 71-83.

[20] CONLEY C.C. (1978) *Isolated Invariant Sets and the Morse Index*, CBMS Reg. Conf. Ser. in Math. 38, AMS, Providence, RI, 1978.

[21] FOSTER D. AND R. VOHRA (1997) Calibrated leaning and correlated equilibria, *Games and Economic Behavior*, **21**, 40-55.

[22] FOSTER D. AND R. VOHRA (1999) Regret in the on-line decision problem, *Games and Economic Behavior*, **29**, 7-35.

[23] FOSTER D. AND P. YOUNG (1998) On the nonconvergence of fictitons play in coordination games, *Games and Economic Behavior*, **25**, 79-96.

[24] FUDENBERG D. AND D. K. LEVINE (1995) Consistency and cautious fictitious play, *Journal of Economic Dynamics and Control*, **19**, 1065-1089.

[25] FUDENBERG D. AND D. K. LEVINE (1998) *The Theory of Learning in Games*, MIT Press.

[26] FUDENBERG D. AND D. K. LEVINE (1999) Conditional universal consistency, *Games and Economic Behavior*, **29**, 104-130.

[27] GAUNERSDORFER A. AND J. HOFBAUER (1995) Fictitious play, Shapley polygons and the replicator equation, *Games and Economic Behavior*, **11**, 279-303.

[28] GILBOA I. AND A. MATSUI (1991) Social stability and equilibrium, *Econometrica*, **59**, 859-867.

[29] HANNAN J. (1957) Approximation to Bayes risk in repeated plays, in Drescher M., A.W. Tucker and P. Wolfe (eds.),*Contributions to the Theory of Games, III*, Princeton University Press, 97-139.

[30] HARRIS C. (1998) On the rate of convergence of continuous time fictitious play, *Games and Economic Behavior*, **22**, 238-259.

[31] HART S. (2005) Adaptive heuristics, *Econometrica*, **73**, 1401-1430.

[32] HART S. AND A. MAS-COLELL (2000) A simple adaptive procedure leading to correlated equilibria, *Econometrica*, **68**, 1127-1150.

[33] HART S. AND A. MAS-COLELL (2003) Regret-based continuous time dynamics, *Games and Economic Behavior*, **45**, 375-394.

[34] HOFBAUER J. (1995) Stability for the best response dynamics, mimeo.

[35] HOFBAUER J. (1998) From Nash and Brown to Maynard Smith: equilibria, dynamics and ESS, *Selection*, **1**, 81-88.

[36] HOFBAUER J. AND W. H. SANDHOLM (2002) On the global convergence of stochastic fictitious play, *Econometrica*, **70**, 2265-2294.

[37] HOFBAUER J. AND W. H. SANDHOLM (2009) Stable games and their dynamics, *Journal of Economic Theory*, **144**, 1665-1693.

[38] HOFBAUER J. AND K. SIGMUND (1998) *Evolutionary Games and Population Dynamics*, Cambridge U.P.

[39] HOFBAUER J. AND K. SIGMUND (2003) Evolutionary games dynamics, *Bulletin A.M.S.*, **40**, 479-519.

[40] HOFBAUER J. AND S. SORIN (2006) Best response dynamics for continuous zero-sum games, *Discrete and Continuous Dynamical Systems-series B*, **6**, 215-224.

[41] HOFBAUER J., S. SORIN AND Y. VIOSSAT (2009) Time average replicator and best reply dynamics, *Mathematics of Operations Research*, **34**, 263-269.

[42] HOPKINS E. (1999) A note on best response dynamics, *Games and Economic Behavior*, **29**, 138-150.

[43] HOPKINS E. (2002) Two competing models of how people learn in games, *Econometrica*, **70**, 2141-2166.

[44] HOPKINS E. AND M. POSCH (2005) Attainability of boundary points under reinforcement learning, *Games and Economic Behavior*, **53**, 110-125.

[45] Krishna V. and T. Sjöstrom (1997) Learning in games: Fictitious play dynamics, in Hart S. and A. Mas-Colell (eds.), *Cooperation: Game-Theoretic Approaches*, NATO ASI Serie A, Springer, 257-273.

[46] Krishna V. and T. Sjöstrom (1988) On the convergence of fictitious play, *Mathematics of Operations Research*, **23**, 479- 511.

[47] Laslier J.-F., R. Topol and B. Walliser (2001) A behavioral learning process in games, *Games and Economic Behavior*, **37**, 340-366.

[48] Leslie D. S. and E.J. Collins, (2005) Individual Q-learning in normal form games, *SIAM Journal of Control and Optimization*, **44**, 495-514.

[49] Littlestone N. and M.K. Warmuth (1994) The weighted majority algorithm, *Information and Computation*, **108**, 212-261.

[50] Maynard Smith J. (1982) *Evolution and the Theory of Games*, Cambridge U.P.

[51] Milgrom P. and J. Roberts (1990) Rationalizability, learning and equilibrium in games with strategic complementarities, *Econometrica*, **58**, 1255-1277.

[52] Milgrom P. and J. Roberts (1991) Adaptive and sophisticated learning in normal form games, *Games and Economic Behavior*, **3**, 82-100.

[53] Monderer D., Samet D and A. Sela (1997) Belief affirming in learning processes, *Journal of Economic Theory*, **73**, 438-452.

[54] Monderer D. and A. Sela (1996) A 2x2 game without the fictitious play property, *Games and Economic Behavior*, **14**, 144-148.

[55] Monderer D. and L.S. Shapley (1996) Potential games, *Games and Economic Behavior*, **14**, 124-143.

[56] Monderer D. and L.S. Shapley (1996) Fictitious play property for games with identical interests, *Journal of Economic Theory*, **68**, 258-265.

[57] Pemantle R. (2007) A survey of random processes with reinforcement, *Probability Surveys*, **4**, 1-79.

[58] Posch M. (1997) Cycling in a stochastic learning algorithm for normal form games, *J. Evol. Econ.*, **7**, 193-207.

[59] Rivière P. (1997) *Quelques Modèles de Jeux d'Evolution*, Thèse, Université P. et M. Curie-Paris 6.

[60] Robinson J. (1951) An iterative method of solving a game, *Annals of Mathematics*, **54**, 296-301.

[61] Shapley L. S. (1964) Some topics in two-person games, in Dresher M., L.S. Shapley and A.W. Tucker (eds.), *Advances in Game Theory*, Annals of Mathematics **52**, Princeton U.P., 1-28.

[62] Sorin S. (2007) Exponential weight algorithm in continuous time, *Mathematical Programming*, Ser. B , **116**, 513-528.

[63] Taylor P. and L. Jonker (1978) Evolutionary stable strategies and game dynamics, *Mathematical Biosciences*, **40**, 145-156.

[64] Viossat Y. (2007) The replicator dynamics does not lead to correlated equilibria, *Games and Economic Behavior*, **59**, 397-407.

[65] Viossat Y. (2008) Evolutionary dynamics may eliminate all strategies used in correlated equilibrium, *Mathematical Social Science*, **56**, 27-43.

[66] Young P. (2004) *Strategic Learning and its Limits*, Oxford U.P. .

Combinatoire et Optimisation, IMJ, CNRS UMR 7586, Faculté de Mathématiques, Université P. et M. Curie - Paris 6, Tour 15-16, 1ière étage, 4 Place Jussieu, 75005 Paris and Laboratoire d'Econométrie, Ecole Polytechnique, France

E-mail address: sorin@math.jussieu.fr

http://www.math.jussieu.fr/ sorin/

Proceedings of Symposia in Applied Mathematics
Volume **69**, 2011

Stochastic Evolutionary Game Dynamics: Foundations, Deterministic Approximation, and Equilibrium Selection

William H. Sandholm

ABSTRACT. We present a general model of stochastic evolution in games played by large populations of anonymous agents. Agents receive opportunities to revise their strategies by way of independent Poisson processes. A *revision protocol* describes how the probabilities with which an agent chooses each of his strategies depend on his current payoff opportunities and the current behavior of the population. Over finite time horizons, the population's behavior is well-approximated by a *mean dynamic*, an ordinary differential equation defined by the expected motion of the stochastic evolutionary process. Over the infinite time horizon, the population's behavior is described by the stationary distribution of the stochastic evolutionary process. If limits are taken in the population size, the level of noise in agents' revision protocols, or both, the stationary distribution may become concentrated on a small set of population states, which are then said to be *stochastically stable*. Stochastic stability analysis allows one to obtain unique predictions of very long run behavior even when the mean dynamic admits multiple locally stable states. We present a full analysis of the asymptotics of the stationary distribution in two-strategy games under noisy best protocols, and discuss extensions of this analysis to other settings.

1. Introduction

Evolutionary game theory studies the behavior of large populations of agents who repeatedly engage in anonymous strategic interactions—that is, interactions in which each agent's outcome depends not only on his own choice, but also on the distribution of others' choices. Applications range from natural selection in animal populations, to driver behavior in highway networks, to consumer choice between different technological standards, to the design of decentralized controlled systems.

In an evolutionary game model, changes in agents' behavior may be driven either by natural selection via differences in birth and death rates in biological contexts, or by the application of myopic decision rules by individual agents in economic contexts. The resulting dynamic models can be studied using tools from the theory of dynamical systems and from the theory of stochastic processes, as

1991 *Mathematics Subject Classification.* Primary 91A22; Secondary 60J20, 37N40.
Key words and phrases. Evolutionary game theory, Markov processes.
I thank Sylvain Sorin for helpful comments. Financial support from NSF Grant SES-0851580 is gratefully acknowledged.

well as those from stochastic approximation theory, which provides important links between the two more basic fields.

In these notes, we present a general model of stochastic evolution in large-population games, and offer a glimpse into the relevant literature by presenting a selection of basic results. In Section 2, we describe population games themselves, and offer a few simple applications. In Sections 3 and 4, we introduce our stochastic evolutionary process. To define this process, we suppose that agents receive opportunities to revise their strategies by way of independent Poisson processes. A *revision protocol* describes how the probabilities with which an agent chooses each of his strategies depend on his current payoff opportunities and the current behavior of the population. Together, a population game, a revision protocol, and a population size implicitly define the stochastic evolutionary process, a Markov process on the set of population states. In Section 4, we show that over finite time horizons, the population's behavior is well-approximated by a *mean dynamic*, an ordinary differential equation defined by the expected motion of the stochastic evolutionary process.

To describe behavior over very long time spans, we turn to an infinite-horizon analysis, in which the population's behavior is described by the stationary distribution of the stochastic evolutionary process. We begin the presentation in Section 5, which reviews the relevant definitions and results from the theory of finite-state Markov processes and presents a number of examples. In order to obtain tight predictions about very long run play, one can examine the limit of the stationary distributions as the population size grows large, the level of noise in agents' decisions becomes small, or both. The stationary distribution may then become concentrated on a small set of population states, which are said to be *stochastically stable.* Stochastic stability analysis allows one to obtain unique predictions of very long run behavior even when the mean dynamic admits multiple locally stable states. In Sections 6 and 7 we introduce the relevant definitions, and we present a full analysis of the asymptotics of the stationary distribution for the case of two-strategy games under noisy best response protocols. This analysis illustrates how the specification of the revision protocol can influence equilibrium selection results. We conclude in Section 8 by discussing extensions of our analyses of infinite-horizon behavior to more complicated strategic settings.

This presentation is based on portions of Chapters 10–12 of Sandholm (2010c), in which a complete treatment of the topics considered here can be found.

2. Population Games

We consider games played by a single population (i.e., games in which all agents play equivalent roles). We suppose that there is a unit mass of agents, each of whom chooses a pure strategy from the set $S = \{1, \dots, n\}$. The aggregate behavior of these agents is described by a *population state*; this is an element of the simplex $X = \{x \in \mathbb{R}^n_+ \colon \sum_{j \in S} x_j = 1\}$, with x_j representing the proportion of agents choosing pure strategy j. We identify a *population game* with a continuous vector-valued payoff function $F : X \to \mathbb{R}^n$. The scalar $F_i(x)$ represents the payoff to strategy i when the population state is x.

Population state x^* is a *Nash equilibrium* of F if no agent can improve his payoff by unilaterally switching strategies. More explicitly, x^* is a Nash equilibrium if

$$x_i^* > 0 \text{ implies that } F_i(x) \geq F_j(x) \text{ for all } j \in S. \tag{1}$$

Example 2.1. In a *symmetric two-player normal form game*, each of the two players chooses a (pure) strategy from the finite set S; which we write generically as $S = \{1, \dots, n\}$. The game's *payoffs* are described by the matrix $A \in \mathbb{R}^{n\times n}$. Entry A_{ij} is the payoff a player obtains when he chooses strategy i and his opponent chooses strategy j; this payoff does not depend on whether the player in question is called player 1 or player 2.

Suppose that the unit mass of agents are randomly matched to play the symmetric normal form game A. At population state x, the (expected) payoff to strategy i is the linear function $F_i(x) = \sum_{j\in S} A_{ij}x_j$; the payoffs to all strategies can be expressed concisely as $F(x) = Ax$. It is easy to verify that x^* is a Nash equilibrium of the population game F if and only if x^* is a symmetric Nash equilibrium of the symmetric normal form game A. ♦

While population games generated by random matching are especially simple, many games that arise in applications are not of this form.

Example 2.2. Consider the following model of highway congestion, due to Beckmann et al. (1956). A pair of towns, Home and Work, are connected by a network of *links*. To commute from Home to Work, an agent must choose a *path* $i \in S$ connecting the two towns. The payoff the agent obtains is the negation of the delay on the path he takes. The delay on the path is the sum of the delays on its constituent links, while the delay on a link is a function of the number of agents who use that link.

Population games embodying this description are known as a *congestion games*. To define a congestion game, let Φ be the collection of links in the highway network. Each strategy $i \in S$ is a route from Home to Work, and so is identified with a set of links $\Phi_i \subseteq \Phi$. Each link ϕ is assigned a *cost function* $c_\phi : \mathbb{R}_+ \to \mathbb{R}$, whose argument is link ϕ's *utilization level* u_ϕ:

$$u_\phi(x) = \sum_{i\in\rho(\phi)} x_i\,, \quad \text{where } \rho(\phi) = \{i \in S : \phi \in \Phi_i\}$$

The payoff of choosing route i is the negation of the total delays on the links in this route:

$$F_i(x) = -\sum_{\phi\in\Phi_i} c_\phi(u_\phi(x)).$$

Since driving on a link increases the delays experienced by other drivers on that link (i.e., since highway congestion involves *negative externalities*), cost functions in models of highway congestion are increasing; they are typically convex as well. Congestion games can also be used to model positive externalities, like the choice between different technological standards; in this case, the cost functions are decreasing in the utilization levels. ♦

3. Revision Protocols and the Stochastic Evolutionary Process

We now introduce foundations for our models of evolutionary dynamics. These foundations are built on the notion of a revision protocol, which describes both the timing and results of agents' myopic decisions about how to continue playing the game at hand. This approach to defining evolutionary dynamics was developed in Björnerstedt and Weibull (1996), Weibull (1995), Hofbauer (1995), and Benaïm and Weibull (2003), and Sandholm (2003, 2010b).

3.1. Definitions. A *revision protocol* is a map $\rho : \mathbb{R}^n \times X \to \mathbb{R}_+^{n\times n}$ that takes the payoff vectors π and population states x as arguments, and returns nonnegative matrices as outputs. For reasons to be made clear below, scalar $\rho_{ij}(\pi, x)$ is called the *conditional switch rate* from strategy i to strategy j.

To move from this notion to an explicit model of evolution, let us consider a population consisting of $N < \infty$ members. A number of the analyses to follow will consider the limit of the present model as the population size N approaches infinity. When the population is of size N, the set of feasible social states is the finite set $\mathcal{X}^N = X \cap \frac{1}{N}\mathbb{Z}^n = \{x \in X : Nx \in \mathbb{Z}^n\}$, a grid embedded in the simplex X.

A revision protocol ρ, a population game F, and a population size N define a continuous-time evolutionary process—a Markov process $\{X_t^N\}$—on the finite state space $\mathcal{X}^N$. A one-size-fits-all description of this process is as follows. Each agent in the society is equipped with a "stochastic alarm clock". The times between rings of of an agent's clock are independent, each with a rate R exponential distribution. The ringing of a clock signals the arrival of a revision opportunity for the clock's owner. If an agent playing strategy $i \in S$ receives a revision opportunity, he switches to strategy $j \neq i$ with probability ρ_{ij}/R. If a switch occurs, the population state changes accordingly, from the old state x to a new state y that accounts for the agent's change in strategy.

To describe the stochastic evolutionary process $\{X_t^N\}$ formally, it is enough to specify its jump rates $\{\lambda_x^N\}_{x\in\mathcal{X}^N}$, which describe the exponential rates of transitions from each state, and its transition probabilities $\{P_{xy}^N\}_{x,y\in\mathcal{X}^N}$, which describe the probabilities that a transition starting at state x ends at state y.

If the current social state is $x \in \mathcal{X}^N$, then Nx_i of the N agents are playing strategy $i \in S$. Since agents receive revision opportunities independently at exponential rate R, the basic properties of the exponential distribution imply that revision opportunities arrive in the society as a whole at exponential rate NR.

When an agent playing strategy $i \in S$ receives a revision opportunity, he switches to strategy $j \neq i$ with probability ρ_{ij}/R. Since this choice is independent of the arrivals of revision opportunities, the probability that the next revision opportunity goes to an agent playing strategy i who then switches to strategy j is

$$\frac{Nx_i}{N} \times \frac{\rho_{ij}}{R} = \frac{x_i\rho_{ij}}{R}.$$

This switch decreases the number of agents playing strategy i by one and increases the number playing j by one, shifting the state by $\frac{1}{N}(e_j - e_i)$.

Summarizing this analysis yields the following observation.

OBSERVATION 3.1. *A population game F, a revision protocol ρ, a constant R, and a population size N define a Markov process $\{X_t^N\}$ on the state space $\mathcal{X}^N$. This process is described by some initial state $X_0^N = x_0^N$, the jump rates $\lambda_x^N = NR$, and the transition probabilities*

$$P_{x\,x+z}^N = \begin{cases} \dfrac{x_i\rho_{ij}(F(x),x)}{R} & \text{if } z = \frac{1}{N}(e_j - e_i),\ i,j \in S,\ i \neq j, \\ 1 - \displaystyle\sum_{i\in S}\sum_{j\neq i} \frac{x_i\rho_{ij}(F(x),x)}{R} & \text{if } z = \mathbf{0}, \\ 0 & \text{otherwise.} \end{cases}$$

3.2. Examples. In economic contexts, revision protocols of the form

$$\rho_{ij}(\pi, x) = x_j r_{ij}(\pi, x) \tag{2}$$

are called *imitative protocols*. These protocols can be given a very simple interpretation: when an agent receives a revision opportunity, he chooses an opponent at random and observes her strategy. If our agent is playing strategy i and the opponent strategy j, the agent switches from i to j with probability proportional to r_{ij}. Notice that the value of the population share x_j is not something the agent need know; this term in (2) accounts for the agent's observing a randomly chosen opponent.

Example 3.2. Suppose that after selecting an opponent, the agent imitates the opponent only if the opponent's payoff is higher than his own, doing so in this case with probability proportional to the payoff difference:

$$\rho_{ij}(\pi, x) = x_j[\pi_j - \pi_i]_+.$$

This protocol is known as *pairwise proportional imitation*; see Helbing (1992) and Schlag (1998). ♦

Additional references on imitative protocols include Björnerstedt and Weibull (1996), Weibull (1995), and Hofbauer (1995).

Protocols of form (2) also appear in biological contexts, starting with the work of Moran (1962), and revisited more recently by Nowak et al. (2004) among others, see Nowak (2006) and Traulsen and Hauert (2009) for further references. In these cases we refer to (2) as a *natural selection protocol*. The biological interpretation of (2) supposes that each agent is programmed to play a single pure strategy. An agent who receives a revision opportunity dies, and is replaced through asexual reproduction. The reproducing agent is a strategy j player with probability $\rho_{ij}(\pi, x) = x_j\hat{\rho}_{ij}(\pi, x)$, which is proportional both to the number of strategy j players and to some function of the prevalences and fitnesses of all strategies. Note that this interpretation requires the restriction

$$\sum_{j\in S} \rho_{ij}(\pi, x) \equiv 1.$$

Example 3.3. Suppose that payoffs are always positive, and let

$$\rho_{ij}(\pi, x) = \frac{x_j \pi_j}{\sum_{k\in S} x_k \pi_k}. \tag{3}$$

Understood as a natural selection protocol, (3) says that the probability that the reproducing agent is a strategy j player is proportional to $x_j\pi_j$, the aggregate fitness of strategy j players.

In economic contexts, we can interpret (3) as an imitative protocol based on repeated sampling. When an agent's clock rings he chooses an opponent at random. If the opponent is playing strategy j, the agent imitates him with probability proportional to π_j. If the agent does not imitate this opponent, he draws a new opponent at random and repeats the procedure. ♦

In the previous examples, only strategies currently in use have any chance of being chosen by a revising agent (or of being the programmed strategy of the newborn agent). Under other protocols, agents' choices are not mediated through the population's current behavior, except indirectly via the effect of behavior on payoffs. These *direct protocols* require agents to directly evaluate the payoffs of each strategy, rather than to indirectly evaluate them as under an imitative procedure.

Example 3.4. Suppose that choices are made according to the *logit choice rule*:

$$\rho_{ij}(\pi, x) = \frac{\exp(\eta^{-1}\pi_j)}{\sum_{k\in S}\exp(\eta^{-1}\pi_k)}. \tag{4}$$

The interpretation of this protocol is simple. Revision opportunities arrive at unit rate. When an opportunity is received by an i player, he switches to strategy j with probability $\rho_{ij}(\pi, x)$, which is proportional to an exponential function of strategy j's payoffs. The parameter $\eta > 0$ is called the *noise level.* If η is large, choice probabilities under the logit rule are nearly uniform. But if η is near zero, choices are optimal with probability close to one, at least when the difference between the best and second best payoff is not too small. ♦

4. Finite Horizon Deterministic Approximation

4.1. Mean Dynamics. A revision protocol ρ, a population game F, and a population size N define a Markov process $\{X_t^N\}$ on the finite state space $\mathcal{X}^N$. We now derive a deterministic process—the *mean dynamic*—that describes the expected motion of $\{X_t^N\}$. In Section 4.3, we will describe formally the sense in which this deterministic process provides a very good approximation of the behavior of the stochastic process $\{X_t^N\}$, at least over finite time horizons and for large population sizes. But having noted this result, we will focus in this section on the deterministic process itself.

To compute the expected increment of $\{X_t^N\}$ over the next dt time units, recall first that each of the N agents receives revision opportunities via a rate R exponential distribution, and so expects to receive $R\,dt$ opportunities during the next dt time units. If the current state is x, the expected number of revision opportunities received by agents currently playing strategy i is approximately $Nx_i\,R\,dt$. Since an i player who receives a revision opportunity switches to strategy j with probability ρ_{ij}/R, the expected number of such switches during the next dt time units is approximately $Nx_i\,\rho_{ij}\,dt$. Therefore, the expected change in the number of agents choosing strategy i during the next dt time units is approximately

$$N\left(\sum_{j\in S} x_j\rho_{ji}(F(x), x) - x_i\sum_{j\in S}\rho_{ij}(F(x), x)\right) dt. \tag{5}$$

Dividing expression (5) by N and eliminating the time differential dt yields a differential equation for the rate of change in the *proportion* of agents choosing strategy i:

$$\dot{x}_i = \sum_{j\in S} x_j\rho_{ji}(F(x), x) - x_i\sum_{j\in S}\rho_{ij}(F(x), x). \tag{M}$$

Equation (M) is the *mean dynamic* (or *mean field*) generated by revision protocol ρ in population game F. The first term in (M) captures the inflow of agents to strategy i from other strategies, while the second captures the outflow of agents to other strategies from strategy i.

4.2. Examples. We now revisit the revision protocols from Section 3.2. To do so, we let

$$\overline{F}(x) = \sum_{i \in S} x_i F_i(x)$$

denote the *average payoff* obtained by the members of the population, and define the *excess payoff* to strategy i,

$$\hat{F}_i(x) = F_i(x) - \overline{F}(x),$$

to be the difference between strategy i's payoff and the population's average payoff.

Example 4.1. In Example 3.2, we introduced the pairwise proportional imitation protocol $\rho_{ij}(\pi, x) = x_j[\pi_j - \pi_i]_+$. This protocol generates the mean dynamic

$$\dot{x}_i = x_i \hat{F}_i(x). \tag{6}$$

Equation (6) is the *replicator dynamic* of Taylor and Jonker (1978), the best-known dynamic in evolutionary game theory. Under this dynamic, the percentage growth rate $\dot{x}_i / x_i$ of each strategy currently in use is equal to that strategy's current excess payoff; unused strategies always remain so. There are a variety of revision protocols other than pairwise proportional imitation that generate the replicator dynamic as their mean dynamics; see Björnerstedt and Weibull (1996) and Hofbauer (1995). ♦

Example 4.2. In Example 3.3, we assumed that payoffs are always positive, and introduced the protocol $\rho_{ij}(\pi, x) \propto x_j \pi_j$, which we interpreted both as a model of biological natural selection and as a model of imitation with repeated sampling. The resulting mean dynamic,

$$\dot{x}_i = \frac{x_i F_i(x)}{\sum_{k \in S} x_k F_k(x)} - x_i = \frac{x_i \hat{F}_i(x)}{\overline{F}(x)}, \tag{7}$$

is the *Maynard Smith replicator dynamic*, due to Maynard Smith (1982). This dynamic only differs from the standard replicator dynamic (6) by a change of speed, with motion under (7) being relatively fast when average payoffs are relatively low. In multipopulation models, the two dynamics are less similar because the changes in speed may differ across populations, affecting the direction of motion. ♦

Example 4.3. In Example 3.4 we introduced the logit choice rule $\rho_{ij}(\pi, x) \propto \exp(\eta^{-1}\pi_j)$. The corresponding mean dynamic,

$$\dot{x}_i = \frac{\exp(\eta^{-1} F_i(x))}{\sum_{k \in S} \exp(\eta^{-1} F_k(x))} - x_i, \tag{8}$$

is called the *logit dynamic*, due to Fudenberg and Levine (1998). ♦

We summarize these and other examples of revision protocols and mean dynamics in Table 1. Dynamics from the table that have not been mentioned so far include the best response dynamic of Gilboa and Matsui (1991), the BNN dynamic of Brown and von Neumann (1950), and the Smith (1984) dynamic. Discussion, examples, and results concerning these and other deterministic dynamics can be found in J. Hofbauer's contribution to this volume.

Revision protocol	Mean dynamic	Name
$\rho_{ij} = x_j[\pi_j - \pi_i]_+$	$\dot{x}_i = x_i \hat{F}_i(x)$	replicator
$\rho_{ij} = \dfrac{\exp(\eta^{-1}\pi_j)}{\sum_{k \in S} \exp(\eta^{-1}\pi_k)}$	$\dot{x}_i = \dfrac{\exp(\eta^{-1}F_i(x))}{\sum_{k \in S} \exp(\eta^{-1}F_k(x))} - x_i$	logit
$\rho_{ij} = 1_{\{j = \text{argmax}_{k \in S} \pi_k\}}$	$\dot{x} \in B^F(x) - x$	best response
$\rho_{ij} = [\pi_j - \sum_{k \in S} x_k \pi_k]_+$	$\dot{x}_i = [\hat{F}_i(x)]_+ - x_i \sum_{j \in S} [\hat{F}_j(x)]_+$	BNN
$\rho_{ij} = [\pi_j - \pi_i]_+$	$\dot{x}_i = \sum_{j \in S} x_j [F_i(x) - F_j(x)]_+ - x_i \sum_{j \in S} [F_j(x) - F_i(x)]_+$	Smith

TABLE 1. Five basic deterministic dynamics.

4.3. Deterministic Approximation Theorem. In Section 3, we defined the Markovian evolutionary process $\{X_t^N\}$ from a revision protocol ρ, a population game F, and a finite population size N. In Section 4.1, we argued that the expected motion of this process is captured by the mean dynamic

$$\text{(M)} \qquad \dot{x}_i = V^F(x) = \sum_{j \in S} x_j \rho_{ji}(F(x), x) - x_i \sum_{j \in S} \rho_{ij}(F(x), x).$$

The basic link between the Markov process $\{X_t^N\}$ and its mean dynamic (M) is provided by the following theorem (Kurtz (1970), Sandholm (2003), Benaïm and Weibull (2003)).

THEOREM 4.4 (Deterministic Approximation of $\{X_t^N\}$). *Suppose that V^F is Lipschitz continuous. Let the initial conditions $X_0^N = x_0^N$ converge to state $x_0 \in X$, and let $\{x_t\}_{t \geq 0}$ be the solution to the mean dynamic* (M) *starting from x_0. Then for all $T < \infty$ and $\varepsilon > 0$,*

$$\lim_{N \to \infty} \mathbb{P}\left(\sup_{t \in [0,T]} \left| X_t^N - x_t \right| < \varepsilon \right) = 1.$$

Thus, when the population size N is large, nearly all sample paths of the Markov process $\{X_t^N\}$ stay within ε of a solution of the mean dynamic (M) through time T. By choosing N large enough, we can ensure that with probability close to one, X_t^N and x_t differ by no more than ε for all t between 0 and T (Figure 1).

The intuition for this result comes from the law of large numbers. At each revision opportunity, the increment in the process $\{X_t^N\}$ is stochastic. Still, the expected number of revision opportunities that arrive during the brief time interval $I = [t, t + dt]$ is large—in particular, of order $N\,dt$. Since each opportunity leads to an increment of the state of size $\frac{1}{N}$, the size of the overall change in the state during time interval I is of order dt. Thus, during this interval there are a large number of revision opportunities, each following nearly the same transition probabilities, and hence having nearly the same expected increments. The law of large numbers therefore suggests that the change in $\{X_t^N\}$ during this interval should be almost completely determined by the expected motion of $\{X_t^N\}$, as described by the mean dynamic (M).

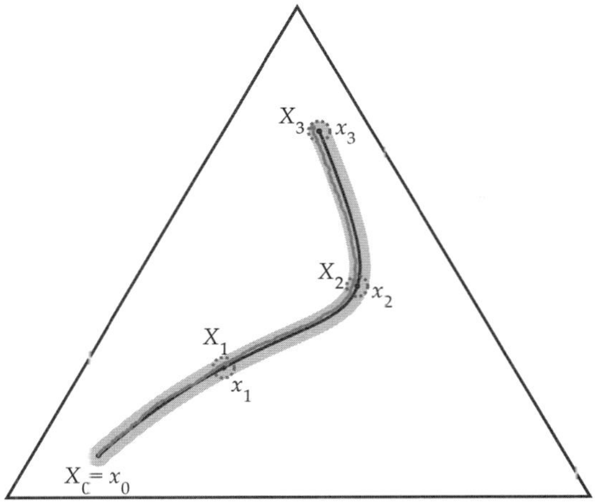

FIGURE 1. Deterministic approximation of the Markov process $\{X_t^N\}$.

It should be emphasized that Theorem 4.4 cannot be extended to an infinite horizon result. To see why not, consider the logit choice protocol (Example 4.3), under which switches between all pairs of strategies occur with positive probability regardless of the current state. As we discuss in Section 5, this property implies that the induced Markov process $\{X_t^N\}$ is irreducible, and hence that every state in $\mathcal{X}^N$ is visited infinitely often with probability one. This fact clearly precludes an infinite horizon analogue of Theorem 4.4. However, the failure of this result introduces a new possibility, that of obtaining unique predictions of infinite horizon behavior. We consider this question in Sections 6 and 7.

4.4. Analysis of Deterministic Dynamics. With this justification in hand, one can use methods from dynamical systems theory to study the behavior of the mean dynamic (M). A large literature has considered this question for a wide range of choices of the revision protocol ρ and the game F, proving a variety of results about local stability of equilibrium, global convergence to equilibrium, and nonconvergence. Other contributions to this volume, in particular those of J. Hofbauer and R. Cressman, address such results; for general references, see Hofbauer and Sigmund (1988, 1998, 2003), Weibull (1995), Sandholm (2009), and chapters 4–9 of Sandholm (2010c).

5. Stationary Distributions

Theorem 4.4 shows that over finite time spans, the stochastic evolutionary process $\{X_t^N\}$ follows a nearly deterministic path, closely shadowing a solution trajectory of the corresponding mean dynamic (M). But if we look at longer time spans—that is, if we fix the population size N of interest and consider the position of the process at large values of t—the random nature of the process must assert itself. If the process is generated by a *full support revision protocol*, one that always assigns positive probabilities to transitions to all neighboring states in $\mathcal{X}^N$, then $\{X_t^N\}$ must visit all states in $\mathcal{X}^N$ infinitely often. Evidently, an *infinite* horizon approximation theorem along the lines of Theorem 4.4 cannot hold. To make predictions about play over very long time spans, we need new techniques for characterizing the infinite horizon behavior of the stochastic evolutionary process. We do so by considering the *stationary distribution* μ^N of the process $\{X_t^N\}$. A stationary distribution is defined by the property that a process whose initial condition is described by this distribution will continue to be described by this distribution at all future times. If $\{X_t^N\}$ is generated by a full support revision protocol, then its stationary distribution μ^N is not only unique, but also describes the infinite horizon behavior of $\{X_t^N\}$ regardless of this process's initial distribution. In principle, this fact allows us to use the stationary distribution to form predictions about a population's very long run behavior that do not depend on its initial behavior. This contrasts sharply with predictions based on the mean dynamic (M), which generally require knowledge of the initial state.

5.1 Full Support Revision Protocols. To introduce the possibility of unique infinite-horizon predictions, we now assume in addition that the conditional switch rates are bounded away from zero: there is a positive constant $\underline{R}$ such that

$$\rho_{ij}(F(x), x) \geq \underline{R} \text{ for all } i, j \in S \text{ and } x \in X. \tag{9}$$

We refer to a revision protocol that satisfies condition (9) as having *full support.*

Example 5.1. Best response with mutations. Under *best response with mutations* at *mutation rate* $\varepsilon > 0$, called *BRM*(ε) for short, a revising agent switches to his current best response with probability $1 - \varepsilon$, but chooses a strategy uniformly at random (or *mutates*) with probability $\varepsilon > 0$. Thus, if the game has two strategies, each yielding different payoffs, a revising agent will choose the optimal strategy with probability $1 - \frac{\varepsilon}{2}$ and will choose the suboptimal strategy with probability $\frac{\varepsilon}{2}$. (Kandori et al. (1993), Young (1993)) ♦

Example 5.2. Logit choice. In Example 3.4 we introduced the logit choice protocol with noise level $\eta > 0$. Here we rewrite this protocol as

$$\rho_{ij}(\pi) = \frac{\exp(\eta^{-1}(\pi_j - \pi_{k^*}))}{\sum_{k \in S} \exp(\eta^{-1}(\pi_k - \pi_{k^*}))}. \tag{10}$$

where k^* is an optimal strategy under π. Then as η approaches zero, the denominator of (10) converges to a constant (namely, the number of optimal strategies under π), so as η^{-1} approaches infinity, $\rho_{ij}(\pi, x)$ vanishes at exponential rate $\pi_{k^*} - \pi_j$. (Blume (1993, 1997)) ♦

As their noise parameters approach zero, both the BRM and logit protocols come to resemble an exact best response protocol. But this similarity masks a fundamental qualitative difference between the two protocols. Under best response with mutations, the probability of choosing a particular suboptimal strategy is independent of the payoff consequences of doing so: mutations do not favor alternative strategies with higher payoffs over those with lower payoffs. In contrast, since the logit protocol is defined using payoff perturbations that are symmetric across strategies, more costly "mistakes" are less likely to be made. One might expect the precise specification of mistake probabilities to be of little consequence. But as we shall see below, predictions of infinite horizon behavior hinge on the relative probabilities of rare events, so that seemingly minor differences in choice probabilities can lead to entirely different predictions of behavior.

5.2. Review: Irreducible Markov Processes. The full support assumption (9) ensures that at each revision opportunity, every strategy in S has a positive probability of being chosen by the revising agent. Therefore, there is a positive probability that the process $\{X_t^N\}$ will transit from any given current state x to any other given state y within a finite number of periods. A Markov process with this property is said to be *irreducible*. Below we review some basic results about infinite horizon behavior of irreducible Markov processes on a finite state space; for details, see e.g. Norris (1997).

Suppose that $\{X_t\}_{t\geq 0}$ is an irreducible Markov process on the finite state space $\mathcal{X}$, where the process has equal jump rates $\lambda_x \equiv l$ and transition matrix P. Then there is a unique probability vector $\mu \in \mathbb{R}_+^{\mathcal{X}}$ satisfying

(11) $$\sum_{x\in\mathcal{X}} \mu_x P_{xy} = \mu_y \quad \text{for all } y \in \mathcal{X}.$$

The vector μ is called the *stationary distribution* of the process $\{X_t\}$. Equation (11) tells us that if we run the process $\{X_t\}$ from initial distribution μ, then at the random time of the first jump, the distribution of the process is also μ. Moreover, if we use the notation $\mathbb{P}_\pi(\cdot)$ to represent $\{X_t\}$ being run from initial distribution π, then

(12) $$\mathbb{P}_\mu\left(X_t = x\right) = \mu_x \quad \text{for all } x \in \mathcal{X} \text{ and } t \geq 0.$$

In other words, if the process starts off in its stationary distribution, it remains in this distribution at all subsequent times t.

While equation (12) tells us what happens if $\{X_t\}$ starts off in its stationary distribution, our main interest is in what happens to this process in the very long run if it starts in an arbitrary initial distribution π. Then as t grows large, the time t distribution of $\{X_t\}$ converges to μ:

(13) $$\lim_{t\to\infty} \mathbb{P}_\pi(X_t = x) = \mu_x \quad \text{for all } x \in \mathcal{X}.$$

Thus, looking at the process $\{X_t\}$ from the ex ante point of view, the probable locations of the process at sufficiently distant future times are essentially determined by μ.

To describe long run behavior from an ex post point of view, we need to consider the behavior of the process's sample paths. Here again, the stationary distribution plays the central role. Then along almost every sample path, the proportion of time spent at each state in the long run is described by μ:

(14) $$\mathbb{P}_\pi\left(\lim_{T\to\infty} \frac{1}{T}\int_0^T 1_{\{X_t=x\}}\,\mathrm{d}t = \mu_x\right) = 1 \quad \text{for all } x \in \mathcal{X}.$$

We can also summarize equation (14) by saying that the limiting empirical distribution of $\{X_t\}$ is almost surely equal to μ.

In general, computing the stationary distribution of a Markov process means finding an eigenvector of a matrix, a task that is computationally daunting unless the state space, and hence the dimension of the matrix, is small. But there is a special class of Markov processes whose stationary distributions are easy to compute. A constant jump rate Markov process $\{X_t\}$ is said to be *reversible* if it admits a *reversible distribution*: a probability distribution μ on $\mathcal{X}$ that satisfies the *detailed balance conditions*:

(15) $$\mu_x P_{xy} = \mu_y P_{yx} \text{ for all } x, y \in \mathcal{X}.$$

A process satisfying this condition is called reversible because, probabilistically speaking, it "looks the same" whether time is run forward or backward. Since summing the equality in (15) over x yields condition (11), a reversible distribution is also a stationary distribution.

While in general reversible Markov processes are rather special, we now introduce one important case in which reversibility is ensured. A constant jump rate Markov process $\{X_t^N\}$ on the state space $\mathcal{X}^N = \{0, \frac{1}{N}, \ldots, 1\}$ is a *birth and death process* if the only positive probability transitions move one step to the right, move one step to the left, or remain still. This implies that there are vectors $p^N, q^N \in \mathbb{R}^{\mathcal{X}^N}$ with $p_1^N = q_0^N = 0$ such that the transition matrix of $\{X_t^N\}$ takes the form

$$P_{\chi y}^N \equiv \begin{cases} p_\chi^N & \text{if } y = \chi + \frac{1}{N}, \\ q_\chi^N & \text{if } y = \chi - \frac{1}{N}, \\ 1 - p_\chi^N - q_\chi^N & \text{if } y = \chi, \\ 0 & \text{otherwise.} \end{cases}$$

Clearly, the process $\{X_t^N\}$ is irreducible if $p_\chi^N > 0$ for $\chi < 1$ and $q_\chi^N > 0$ for $\chi > 0$, as we henceforth assume. For the transition matrix above, the reversibility conditions (15) reduce to

$$\mu_\chi^N q_\chi^N = \mu_{\chi-1/N}^N \, p_{\chi-1/N}^N \quad \text{for } \chi \in \{\tfrac{1}{N}, \ldots, 1\}.$$

Applying this formula inductively, we find that the stationary distribution of $\{X_t^N\}$ satisfies

(16) $$\frac{\mu_\chi^N}{\mu_0^N} = \prod_{j=1}^{N\chi} \frac{p_{(j-1)/N}^N}{q_{j/N}^N} \quad \text{for } \chi \in \{\tfrac{1}{N}, \ldots, 1\},$$

with μ_0 determined by the requirement that the weights in μ^N must sum to 1.

5.3. Stationary Distributions for Two-Strategy Games. When the population plays a game with just two strategies, the state space $\mathcal{X}^N$ is a grid in the simplex in $\mathbb{R}^2$. In this case it is convenient to identify state x with the weight $\chi \equiv x_1$ that it places on strategy 1. Under this notational device, the state space of the Markov process $\{X_t^N\}$ becomes $\mathcal{X}^N = \{0, \frac{1}{N}, \ldots, 1\}$, a uniformly-spaced grid in the unit interval. We will also write $F(\chi)$ for $F(x)$ and $\rho(\pi, \chi)$ for $\rho(\pi, x)$ whenever it is convenient to do so.

Because agents in our model switch strategies sequentially, transitions of the process $\{X_t^N\}$ are always between adjacent states, implying that $\{X_t^N\}$ is a birth and death processes. Let us now use formula (16) to compute the stationary distribution of our stochastic evolutionary process, maintaining the assumption that the process is generated by a full support revision protocol. Referring back to Section 3.1, we find that the process $\{X_t^N\}$ has constant jump rates $\lambda_\chi^N = NR$, and that its upward and downward transition probabilities are given by

(17) $$p_\chi^N = (1 - \chi) \cdot \tfrac{1}{R}\rho_{01}(F(\chi), \chi) \text{ and}$$

(18) $$q_\chi^N = \chi \cdot \tfrac{1}{R}\rho_{10}(F(\chi), \chi).$$

Substituting formulas (17) and (18) into equation (16), we see that for $\chi \in \{\frac{1}{N}, \frac{2}{N}, \ldots, 1\}$, we have

$$\frac{\mu_\chi^N}{\mu_0^N} = \prod_{j=1}^{N\chi} \frac{p_{(j-1)/N}^N}{q_{j/N}^N} = \prod_{j=1}^{N\chi} \frac{(1 - \frac{j-1}{N})}{\frac{j}{N}} \cdot \frac{\frac{1}{R}\rho_{01}(F(\frac{j-1}{N}), \frac{j-1}{N})}{\frac{1}{R}\rho_{10}(F(\frac{j}{N}), \frac{j}{N})}.$$

Simplifying this expression yields the following result.

THEOREM 5.3. *Suppose that a population of N agents plays the two-strategy game F using the full support revision protocol ρ. Then the stationary distribution for the evolutionary process $\{X_t^N\}$ on $\mathcal{X}^N$ is given by*

$$\frac{\mu_\chi^N}{\mu_0^N} = \prod_{j=1}^{N\chi} \frac{(N - j + 1)}{j} \cdot \frac{\rho_{01}(F(\frac{j-1}{N}), \frac{j-1}{N})}{\rho_{10}(F(\frac{j}{N}), \frac{j}{N})} \quad \text{for } \chi \in \{\tfrac{1}{N}, \tfrac{2}{N}, \ldots, 1\},$$

with μ_0^N determined by the requirement that $\sum_{\chi \in \mathcal{X}^N} \mu_\chi^N = 1$.

In what follows, we will use Theorem 5.3 to understand the infinite-horizon behavior of the process $\{X_t^N\}$, in particular as various parameters are taken to their limiting values.

5.4. Examples. The power of infinite horizon analysis lies in its ability to generate unique predictions of play even in games with multiple strict equilibria. We now illustrate this idea by computing some stationary distributions for two-strategy coordination games under the BRM and logit rules. In all cases, we find that these distributions place most of their weight near a single equilibrium. But we also find that the two rules need not select the same equilibrium.

Example 5.4. Stag Hunt. The symmetric normal form coordination game

$$A = \begin{pmatrix} h & h \\ 0 & s \end{pmatrix}$$

with $s > h > 0$ is known as *Stag Hunt*. By way of interpretation, we imagine that each agent in a match must decide whether to hunt for hare or for stag. Hunting for hare ensures a payoff of h regardless of the match partner's choice. Hunting for stag can generate a payoff of $s > h$ if the opponent does the same, but results in a zero payoff otherwise. Each of the two strategies has distinct merits. Coordinating on Stag yields higher payoffs than coordinating on Hare. But the payoff to Hare is certain, while the payoff to Stag depends on the choice of one's partner.

Suppose that a population of agents is repeatedly matched to play Stag Hunt. If we let χ denote the proportion of agents playing Stag, then with our usual abuse of notation, the payoffs in the resulting population game are $F_H(\chi) = h$ and $F_S(\chi) = s\chi$. This population game has three Nash equilibria: the two pure equilibria, and the mixed equilibrium $\chi^* = \frac{h}{s}$. We henceforth suppose that $h = 2$ and $s = 3$, so that the mixed equilibrium places mass $\chi^* = \frac{2}{3}$ on Stag.

Suppose that agents follow the best response with mutations protocol, with mutation rate $\varepsilon = .10$. The resulting mean dynamic,

$$\dot{\chi} = \begin{cases} \frac{\varepsilon}{2} - \chi & \text{if } \chi < \frac{2}{3}, \\ (1 - \frac{\varepsilon}{2}) - \chi & \text{if } \chi > \frac{2}{3}, \end{cases}$$

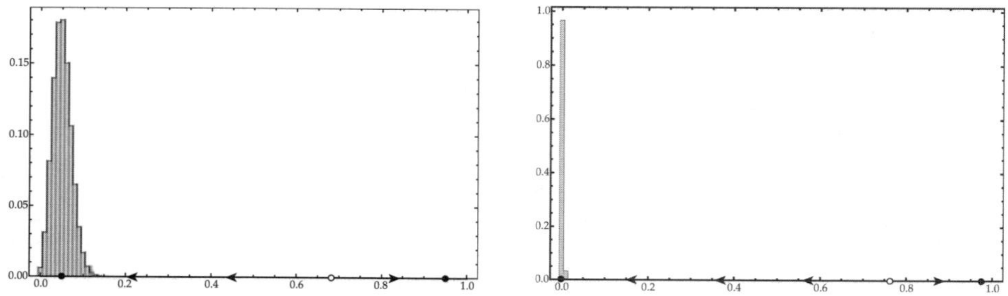

(I) best response with mutations ($\varepsilon = .10$)

(II) logit ($\eta = .25$)

FIGURE 2. Stationary distribution weights μ_χ for Stag Hunt ($h = 2$, $s = 3$, $N = 100$).

has stable rest points at $\chi = .05$ and $\chi = .95$. The basins of attraction of these rest points meet at the mixed equilibrium $\chi^* = \frac{2}{3}$. Note that the rest point that approximates the all-Hare equilibrium has the larger basin of attraction.

In Figure 2(i), we present this mean dynamic underneath the stationary distribution μ^N for $N = 100$, which we computed using the formula derived in Theorem 5.3. While the mean dynamic has two stable equilibria, nearly all of the mass in the stationary distribution is concentrated at states where between 88 and 100 agents choose Hare. Thus, while coordinating on Stag is efficient, the "safe" strategy Hare is selected by the stochastic evolutionary process.

Suppose instead that agents use the logit rule with noise level $\eta = .25$. The mean dynamic is then the logit dynamic,

$$\dot{\chi} = \frac{\exp(3\chi\eta^{-1})}{\exp(2\eta^{-1}) + \exp(3\chi\eta^{-1})} - \chi,$$

which has stable rest points at $\chi = .0003$ and $\chi = .9762$, and an unstable rest point at $\chi = .7650$, so that the basin of attraction of the "almost all-Hare" rest point $\chi = .0003$ is even larger than under BRM. Examining the resulting stationary distribution (Figure 2(ii)), we see that virtually all of its mass is placed on states where either 99 or 100 agents choose Hare, in rough agreement with the result for the BRM(.10) rule. ♦

Why does most of the mass in the stationary distribution becomes concentrated around a single equilibrium? The stochastic evolutionary process $\{X_t^N\}$ typically moves in the direction indicated by the mean dynamic. If the process begins in the basin of attraction of a rest point or other attractor of this dynamic, then the initial period of evolution generally results in convergence to and lingering near this locally stable set.

However, since BRM and logit choice lead to irreducible evolutionary processes, this cannot be the end of the story. Indeed, we know that the process $\{X_t^N\}$ eventually reaches all states in $\mathcal{X}^N$; in fact, it visits all states in $\mathcal{X}^N$ infinitely often. This means that the process at some point must leave the basin of the stable set visited first; it then enters the basin of a new stable set, at which point it is extremely likely to head directly the set itself. The evolution of the process

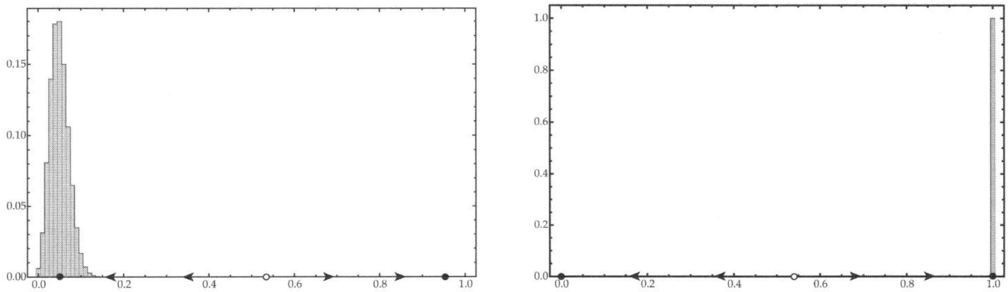

(I) best response with mutations ($\varepsilon = .10$)

(II) logit ($\eta = .25$)

FIGURE 3. Stationary distribution weights μ_χ for a nonlinear Stag Hunt ($h = 2$, $s = 7$, $N = 100$).

continues in this fashion, with long periods of visits to each attractor punctuated by sudden jumps between the stable set.

Which states are visited most often over the infinite horizon is determined by the *relative* unlikelihoods of these rare but inevitable transitions between stable sets. In the examples above, the transitions from the Stag rest point to the Hare rest point and from the Hare rest point to the Stag rest point are both very unlikely events. But for purposes of determining the stationary distribution, what matters is that in relative terms, the former transitions are much more likely than the latter. This enables us to conclude that over very long time spans, the evolutionary process will spend most periods at states where most agents play Hare.

Example 5.5. A nonlinear Stag Hunt. We now consider a version of the Stag Hunt game in which payoffs depend nonlinearly on the population state. With our usual abuse of notation, we define payoffs in this game by $F_H(\chi) = h$ and $F_S(\chi) = s\chi^2$, with χ representing the proportion of agents playing Stag. The population game F has three Nash equilibria: the pure equilibria $\chi = 0$ and $\chi = 1$, and the mixed equilibrium $\chi^* = \sqrt{h/s}$. We focus on the case in which $h = 2$ and $s = 7$, so that $\chi^* = \sqrt{2/7} \approx .5345$.

Suppose first that a population of 100 agents play this game using the BRM(.10) rule. In Figure 3(i) we present the resulting mean dynamic beneath a graph of the stationary distribution μ^{100}. The mean dynamic has rest points at $\chi = .05$, $\chi = .95$, and $\chi^* \approx .5345$, so the the "almost all Hare" rest point again has the larger basin of attraction. As was true in the linear Stag Hunt from Example 5.4, the stationary distribution generated by the BRM(.10) rule in this nonlinear Stag Hunt places nearly all of its mass on states where at least 88 agents choose Hare.

Figure 3(ii) presents the mean dynamic and the stationary distribution μ^{100} for the logit rule with $\eta = .25$. The rest points of the logit(.25) dynamic are $\chi = .0003$, $\chi = 1$, and $\chi = .5398$, so the "almost all Hare" rest point once again has the larger basin of attraction. Nevertheless, the stationary distribution μ^{100} places virtually all of its mass on the state in which all 100 agents choose Stag.

To summarize, our prediction for very long run behavior under the BRM(.10) rule is inefficient coordination on Hare, while our prediction under the logit(.25) rule is efficient coordination on Stag. ♦

For the intuition behind this discrepancy in predictions, recall the discussion from Section 5.1 about the basic distinction between the logit and BRM protocols: under logit choice, the probability of a "mistake" depends on its payoff consequences, while under BRM, it does not. The latter observation implies that under BRM, the probabilities of escaping from the basins of attraction of stable sets, and hence the identities of the states that predominate in the very long run, depend only on the size and the shapes of the basins. In the current one-dimensional example, these shapes are always line segments, so that only the size of the basins matters; since the "almost all-Hare" state has the larger basin, it is selected under the BRM rule.

On the contrary, the probability of escaping a stable equilibrium under logit choice depends not only on the shape and size of its basin, but also on the payoff differences that must be overcome during the journey. In the nonlinear Stag Hunt game, the basin of the "almost all-Stag" equilibrium is smaller than that of the all-Hare equilibrium. But because the payoff advantage of Stag over Hare in the former's basin tends to be much larger than the payoff advantage of Hare over Stag in the latter's, it is more difficult for the population to escape the all-Stag equilibrium than the all-Hare equilibrium; as a result, the population spends virtually all periods coordinating on Stag over the infinite horizon.

We can compare the process of escaping from the basin of a stable rest point to an attempt to swim upstream. Under BRM, the strength of the stream's flow is constant, so the difficulty of a given excursion is proportional to distance. Under logit choice, the strength of the stream's flow is variable, so the difficulty of an excursion depends on how this strength varies over the distance travelled. In general, the probability of escaping from a stable set is determined by both the distance that must be travelled and the strength of the oncoming flow.

To obtain unique predictions of infinite horizon behavior, it is generally enough either that the population size not be too small, or that the noise level in agents' choices not be too large. But one can obtain cleaner and more general results by studying the limiting behavior of the stationary distribution as the population size approaches infinity, the noise level approaches zero, or both. This approach to studying infinite horizon behavior, known as *stochastic stability theory.*

One difficulty that can arise in this setting is that the prediction of infinite horizon behavior can depend on the identity or on the order in which limits are taken. Our last example, based on Binmore and Samuelson (1997), illustrates this point.

Example 5.6. Consider a population of agents who are matched to play the symmetric normal form game with strategy set $S = \{0, 1\}$ and payoff matrix

$$A = \begin{pmatrix} 1 & 2 \\ 3 & 1 \end{pmatrix}.$$

The unique Nash equilibrium of the population game $F(x) = Ax$ is the mixed equilibrium $x^* = (x_0^*, x_1^*) = (\frac{1}{3}, \frac{2}{3})$. To simplify notation in what follows we allow self-matching, but the analysis is virtually identical without it.

Suppose that agents employ the following revision protocol, which combines imitation of successful opponents and mutations:

$$\rho_{ij}^{\varepsilon}(\pi, x) = x_j \pi_j + \varepsilon.$$

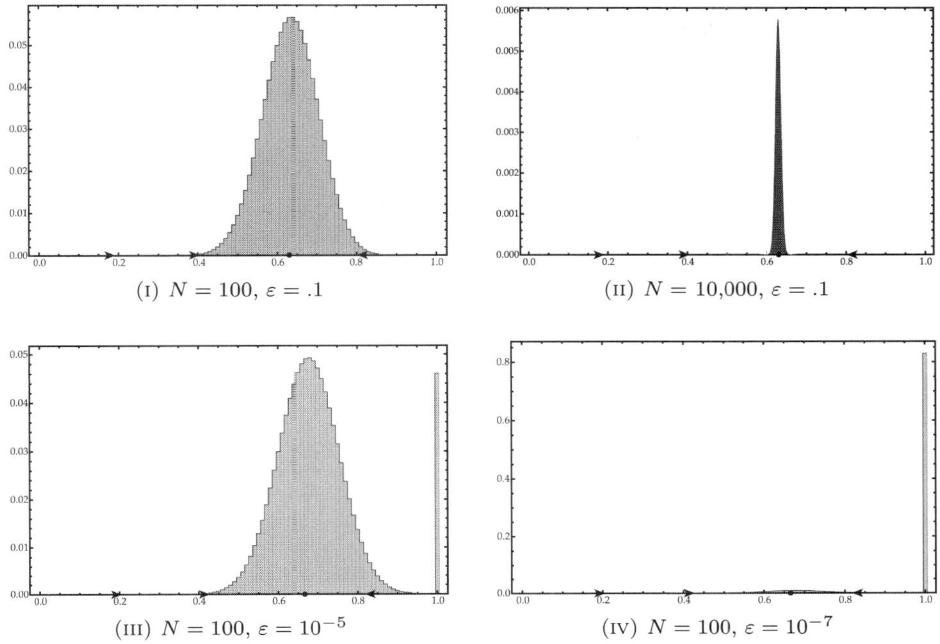

FIGURE 4. Stationary distribution weights $\mu_{\chi}^{N,\varepsilon}$ in an anticoordination game under an "imitation with mutation" protocol.

The protocol ρ^{ε} generates the mean dynamic

$$\dot{x}_i = V_i^{\varepsilon}(x) = x_i \hat{F}_i(x) + 2\varepsilon\left(\tfrac{1}{2} - x_i\right), \tag{19}$$

which is the sum of the replicator dynamic and an order ε term that points toward the center of the simplex. When $\varepsilon = 0$, this dynamic is simply the replicator dynamic: the Nash equilibrium $x^* = (\frac{1}{3}, \frac{2}{3})$ attracts solutions from all interior initial conditions, while pure states e_0 and e_1 are unstable rest points. When $\varepsilon > 0$, the two boundary rest points disappear, leaving a globally stable rest point that is near x^*, but slightly closer to the center of the simplex.

Using the formulas from Theorem 5.3, we can compute the stationary distribution $\mu^{N,\varepsilon}$ of the process $\{X_t^{N,\varepsilon}\}$ generated by F and ρ^{ε} for any fixed values of N and ε. Four instances are presented in Figure 4.

Figure 4(i) presents the stationary distribution when $\varepsilon = .1$ and $N = 100$. This distribution is drawn above the phase diagram of the mean dynamic (19), whose global attractor is appears at $\hat{\chi} \approx .6296$. The stationary distribution $\mu^{N,\varepsilon}$ has its mode at state $\chi = .64$, but is dispersed rather broadly about this state.

Figure 4(ii) presents the stationary distribution and mean dynamic when $\varepsilon = .1$ and $N = 10{,}000$. Increasing population size moves the mode of the distribution occurs to state $\chi = .6300$, and, more importantly, causes the distribution to exhibit much less dispersion around the modal state. This numerical analysis suggests that in the large population limit, the stationary distribution $\mu^{N,\varepsilon}$ will approach a point mass at $\hat{\chi} \approx .6296$, the global attractor of the relevant mean dynamic.

As the noise level ε approaches zero, the rest point of the mean dynamic approaches the Nash equilibrium $\chi^* = \frac{2}{3}$. Therefore, if after taking N to infinity we take ε to zero, we obtain the double limit

$$\lim_{\varepsilon \to 0} \lim_{N \to \infty} \mu^{N,\varepsilon} = \delta_{\chi^*}, \tag{20}$$

where the limits refer to weak convergence of probability measures, and δ_{χ^*} denotes the point mass at state χ^*.

The remaining pictures illustrate the effects of setting very small mutation rates. When $N = 100$ and $\varepsilon = 10^{-5}$ (Figure 4(iii)) most of the mass in $\mu^{100,\varepsilon}$ falls in a bell-shaped distribution centered at state $\chi = .68$, but a mass of $\mu_1^{100,\varepsilon} = .0460$ sits in isolation at the boundary state $\chi = 1$. When ε is reduced to 10^{-7} (Figure 4(iv)), this boundary state commands a majority of the weight in the disribution ($\mu_1^{100,\varepsilon} = .8286$).

This numerical analysis suggests that when the mutation rate approaches zero, the stationary distribution will approach a point mass at state 1. Increasing the population size does not alter this result, so for the small noise double limit we obtain

$$\lim_{N \to \infty} \lim_{\varepsilon \to 0} \mu^{N,\varepsilon} = \delta_1, \tag{21}$$

where δ_1 denotes the unit point mass at state 1.

Comparing equations (20) and (21), we conclude that the large population double limit and the small noise double limit disagree. ♦

In the preceding example, the large population limits agree with the predictions of the mean dynamic, while the small noise limits do not. Still, the behavior of the latter limits is easy to explain. Starting from any interior state, and from the boundary as well when $\varepsilon > 0$, the expected motion of the process $\{X_t^{N,\varepsilon}\}$ is toward the interior rest point of the mean dynamic V^ε. But when ε is zero, the boundary states 0 and 1 become rest points of V^ε, and are absorbing states of $\{X_t^{N,\varepsilon}\}$; in fact, it is easy to see that they are the only recurrent states of the zero-noise process. Therefore, when $\varepsilon = 0$, $\{X_t^{N,\varepsilon}\}$ reaches either state 0 or state 1 in finite time, and then remains at that state forever.

If instead ε is positive, the boundary states are no longer absorbing, and they are far from any rest point of the mean dynamic. But once the process $\{X_t^{N,\varepsilon}\}$ reaches such a state, it can only depart by way of a mutation. Thus, if we fix the population size N and make ε extremely small, then a journey from an interior state to a boundary state—here a journey against the flow of the mean dynamic—is "more likely" than an escape from a boundary state by way of a single mutation. It follows that in the small noise limit, the stationary distribution must become concentrated on the boundary states regardless of the nature of the mean dynamic. (In fact, it will typically become concentrated on just one of these states.)

As this discussion indicates, the prediction provided by the small noise limit does not become a good approximation of behavior at fixed values of N and ε unless ε is so small that lone mutations are much more rare than excursions from the interior of $\mathcal{X}^N$ to the boundary. In Figures 4(iii) and (iv), which consider a modest population size of $N = 100$, we see that a mutation rate of $\varepsilon = 10^{-5}$ is not small enough to yield agreement with the prediction of the small noise limit, though a mutation rate of $\varepsilon = 10^{-7}$ yields a closer match. With larger population sizes, the relevant mutation rates would be even smaller.

This example suggests that in economic contexts, where the probabilities of "mutations" may not be especially small, the large population limit is more likely to be the relevant one in cases where the predictions of the two limits disagree. In biological contexts, where mutation rates may indeed be quite small, the choice between the limits seems less clear.

6. Asymptotics of the Stationary Distribution and Stochastic Stability

The examples in Section 5 show that even when the underlying game has multiple strict equilibria, the stationary distribution is often concentrated in the vicinity of just one of them if the noise level η is small or the population size N is large. In these cases, the population state so selected provides a unique prediction of infinite horizon play.

In order to obtain clean selection results, we now allow the parameters η and N to approach their limiting values. While each fixed stationary distribution $\mu^{N,\eta}$ has full support on $\mathcal{X}^N$, the limit of a sequence of stationary distributions may converge to a point mass at a single state; thus, taking limits in η and N allows us to obtain exact equilibrium selection results. Moreover while computing a particular stationary distribution requires solving a large collection of linear equalities, the limiting stationary distribution can often be found without explicitly computing any of the stationary distributions along the sequence (see Section 8).

Population states that retain mass in a limiting stationary distribution are said to be *stochastically stable*. There are a number of different definitions of stochastic stability, depending on which limits are taken—just η, just N, η followed by N, or N followed by η—and on what should count as "retaining mass". Taking only the small noise limit, or taking this limit first, emphasizes the rarity of suboptimal play as the key force behind equilibrium selection. Taking only the large population limit, or taking it first, emphasizes the effects of large numbers of conditionally independent decisions in driving equilibrium selection. Since it is not always easy to know which of these forces should be viewed as the primary one, a important goal in stochastic stability analysis is to identify settings in which the small noise and large population limits agree.

Analysis of stochastic stability in games have been carried out under a wide array of assumptions about the form of the underlying game, the nature of the revision protocol, the specification of the evolutionary process, and the limits taken to define stochastic stability—see Section 8 and Sandholm (2009, 2010c) for references. In what follows, we focus on an interesting setting in which all calculations can be carried until their very end, and in which one can obtain precise statements about infinite horizon behavior and about the agreement of the small noise and large population limits.

7. Noisy Best Response Protocols in Two-Strategy Games

Here we consider evolution in two-strategy games under a general class of noisy best response protocols. We introduce the notion of the *cost* of a suboptimal choice, which is defined as the rate of decay of the probability of making this choice as the noise level approaches zero. Using this notion, we derive simple formulas that characterize the asymptotics of the stationary distribution under the various limits in η and N, and offer a necessary and sufficient condition for an equilibrium to be uniquely stochastically stable under every noisy best response protocol. This section follows Sandholm (2010a), which builds on earlier work by Binmore and Samuelson (1997), Blume (2003), and Sandholm (2007).

7.1. Noisy Best Response Protocols and their Cost Functions. We consider evolution under *noisy best response protocols*. These protocols can be expressed as

$$\rho^{\eta}_{ij}(\pi) = \sigma^{\eta}(\pi_j - \pi_i), \tag{22}$$

for some function $\sigma^{\eta} \colon \mathbb{R} \to (0,1)$: when a current strategy i player receives a revision opportunity, he switches to strategy $j \neq i$ with a probability that only depends on the payoff advantage of strategy j over strategy i. To justify its name, the protocol σ^{η} should recommend optimal strategies with high probability when the noise level is small:

$$\lim_{\eta \to 0} \sigma^{\eta}(a) = \begin{cases} 1 & \text{if } a > 0, \\ 0 & \text{if } a < 0. \end{cases}$$

To place further structure on the probabilities of suboptimal choices, we impose restrictions on the rates at which the probabilities $\sigma^{\eta}(a)$ of choosing a suboptimal strategy approach zero as η approaches zero. To do so, we define the *cost* of switching to a strategy with payoff *disadvantage* $d \in \mathbb{R}$ as

$$\kappa(d) = -\lim_{\eta \to 0} \eta \log \sigma^{\eta}(-d). \tag{23}$$

By unpacking this expression, we can write the probability of switching to a strategy with payoff disadvantage d when the noise level is η as

$$\sigma^{\eta}(-d) = \exp\left(-\eta^{-1}(\kappa(d) + o(1))\right),$$

where $o(1)$ represents a term that vanishes as η approaches 0. Thus, $\kappa(d)$ is the exponential rate of decay of the choice probability $\sigma^{\eta}(-d)$ as η^{-1} approaches infinity.

We are now ready to define the class of protocols we will consider.

DEFINITION. We say that the noisy best response protocol (22) is *regular* if

(i) the limit in (23) exists for all $d \in \mathbb{R}$, with convergence uniform on compact intervals;
(ii) κ is nondecreasing;
(iii) $\kappa(d) = 0$ whenever $d < 0$;
(iv) $\kappa(d) > 0$ whenever $d > 0$.

Conditions (ii)-(iv) impose constraints on the rates of decay of switching probabilities. Condition (ii) requires the rate of decay to be nondecreasing in the payoff disadvantage of the alternative strategy. Condition (iii) requires the switching probability of an agent currently playing the suboptimal strategy to have rate of decay zero; the condition is satisfied when the probability is bounded away from zero, although this is not necessary for the condition to hold. Finally, condition (iv) requires the probability of switching from the optimal strategy to the suboptimal one to have a positive rate of decay. These conditions are consistent with having either $\kappa(0) > 0$ or $\kappa(0) = 0$: thus, when both strategies earn the same payoff, the probability that a revising agent opts to switch strategies can converge to zero with a positive rate of decay, as in Example 7.1 below, or can be bounded away from zero, as in Examples 7.2 and 7.3.

We now present the three leading examples of noisy best response protocols.

Example 7.1. Best response with mutations. The BRM protocol with noise level $\eta\ (= -(\log \varepsilon)^{-1})$, introduced in Example 5.1, is defined by

$$\sigma^\eta(a) = \begin{cases} 1 - \exp(-\eta^{-1}) & \text{if } a > 0, \\ \exp(-\eta^{-1}) & \text{if } a \leq 0. \end{cases}$$

In this specification, an indifferent agent only switches strategies in the event of a mutation. Since for $d \geq 0$ we have $-\eta \log \sigma^\eta(-d) = 1$, protocol σ^η is regular with cost function

$$\kappa(d) = \begin{cases} 1 & \text{if } d \geq 0, \\ 0 & \text{if } d < 0. \end{cases} \quad \blacklozenge$$

Example 7.2. Logit choice. The *logit choice protocol* with noise level $\eta > 0$, introduced in Examples 3.4 and 5.2, is defined in two-strategy games by

$$\sigma^\eta(a) = \frac{\exp(\eta^{-1}a)}{\exp(\eta^{-1}a) + 1}.$$

For $d \geq 0$, we have that $-\eta \log \sigma^\eta(-d) = d + \eta \log(\exp(-\eta^{-1}d) + 1)$. It follows that σ^η is regular with cost function

$$\kappa(d) = \begin{cases} d & \text{if } d > 0, \\ 0 & \text{if } d \leq 0. \end{cases} \quad \blacklozenge$$

Example 7.3. Probit choice. The logit choice protocol can be derived from a random utility model in which the strategies' payoffs are perturbed by i.i.d., double exponentially distributed random variables (see Hofbauer and Sandholm (2002)). The *probit choice protocol* assumes instead that the payoff perturbations are i.i.d. normal random variables with mean 0 and variance η. Thus

$$\sigma^\eta(a) = \mathbb{P}(\sqrt{\eta}\, Z + a > \sqrt{\eta}\, Z'),$$

where Z and Z' are independent and standard normal. It follows easily that

$$\sigma^\eta(a) = \Phi\left(\tfrac{a}{\sqrt{2\eta}}\right), \tag{24}$$

where Φ is the standard normal distribution function.

A well-known approximation of Φ tells us that when $z < 0$,

$$\Phi(z) = K(z) \exp(\tfrac{-z^2}{2}) \tag{25}$$

for some $K(z) \in (\frac{-1}{\sqrt{2\pi}\, z}(1 - \frac{1}{z^2}), \frac{-1}{\sqrt{2\pi}\, z})$. By employing this observation, one can show that σ^η is regular with cost function

$$\kappa(d) = \begin{cases} \frac{1}{4}d^2 & \text{if } d > 0, \\ 0 & \text{if } d \leq 0. \end{cases} \quad \blacklozenge$$

7.2. The (Double) Limit Theorem. Our result on the asymptotics of the stationary distribution requires a few additional definitions and assumptions. We suppose that the sequence of two-strategy games $\{F^N\}_{N=N_0}^\infty$ converges uniformly to a continuous-population game F, where $F\colon [0,1] \to \mathbb{R}^2$ is a continuous function. We let

$$F_\Delta(\chi) \equiv F_1(\chi) - F_0(\chi)$$

denote the payoff advantage of strategy 1 at state χ in the limit game.

We define the *relative cost function* $\tilde{\kappa} \colon \mathbb{R} \to \mathbb{R}$ by

$$\tilde{\kappa}(d) = \lim_{\eta \to 0} \left(-\eta \log \sigma^{\eta}(-d) + \eta \log \sigma^{\eta}(d) \right) = \kappa(d) - \kappa(-d). \tag{26}$$

Our assumptions on κ imply that $\tilde{\kappa}$ is nondecreasing, sign preserving ($\text{sgn}(\tilde{\kappa}(d)) = \text{sgn}(d)$), and odd ($\tilde{\kappa}(d) = -\tilde{\kappa}(-d)$).

We define the *ordinal potential function* $I : [0,1] \to \mathbb{R}$ by

$$I(\chi) = \int_0^{\chi} \tilde{\kappa}(F_{\Delta}(y))\, dy, \tag{27}$$

where the relative cost function $\tilde{\kappa}$ is defined in equation (26). Observe that by marginally adjusting the state χ so as to increase the mass on the optimal strategy, we increase the value of I at rate $\tilde{\kappa}(a)$, where a is the optimal strategy's payoff advantage. Thus, the ordinal potential function combines information about payoff differences with the costs of the associated suboptimal choices.

Finally, we define $\Delta I \colon [0,1] \to (-\infty, 0]$ by

$$\Delta I(\chi) = I(\chi) - \max_{y \in [0,1]} I(y). \tag{28}$$

Thus, ΔI is obtained from I by shifting its values uniformly, doing so in such a way that the maximum value of ΔI is zero.

Example 7.4. If ρ^{η} represents best response with mutations (Example 7.1), then the ordinal potential function (27) becomes the *signum potential function*

$$I_{\text{sgn}}(\chi) = \int_0^{\chi} \text{sgn}\left(F_{\Delta}(y)\right) dy.$$

This slope of this function at state χ is 1, -1, or 0, according to whether the optimal strategy at χ is strategy 1, strategy 0, or both. ♦

Example 7.5. If ρ^{η} represents logit choice (Example 7.2), then (27) becomes the (*standard*) *potential function*

$$I_1(\chi) = \int_0^{\chi} F_{\Delta}(y)\, dy,$$

whose slope at state χ is just the payoff difference at χ. ♦

Example 7.6. If ρ^{η} represents probit choice (Example 7.3), then (27) becomes the *quadratic potential function*

$$I_2(\chi) = \int_0^{\chi} \tfrac{1}{4} \left\langle F_{\Delta}(y) \right\rangle^2 dy,$$

where $\langle a \rangle^2 = \text{sgn}(a)\, a^2$ is the signed square function. The values of I_2 again depend on payoff differences, but relative to the logit case, larger payoff differences play a more important role. This contrast can be traced to the fact that at small noise levels, the double exponential distribution has fatter tails than the normal distribution—compare Example 7.3. ♦

Theorem 7.7 shows that whether one takes the small noise limit before the large population limit, or the large population before the small noise limit, the rates of decay of the stationary distribution are captured by the ordinal potential function I. Since the double limits agree, our predictions of infinite horizon behavior under noisy best response rules do not depend on which force drives the equilibrium selection results.

THEOREM 7.7. *The stationary distributions $\mu^{N,\eta}$ satisfy*

$$(i) \qquad \lim_{N\to\infty}\lim_{\eta\to 0}\max_{\chi\in\mathcal{X}^N}\left|\tfrac{\eta}{N}\log\mu_\chi^{N,\eta}-\Delta I(\chi)\right|=0 \quad \textit{and}$$

$$(ii) \qquad \lim_{\eta\to 0}\lim_{N\to\infty}\max_{\chi\in\mathcal{X}^N}\left|\tfrac{\eta}{N}\log\mu_\chi^{N,\eta}-\Delta I(\chi)\right|=0.$$

Theorem 7.7 is proved by manipulating the stationary distribution formula from Theorem 5.3 and applying the dominated convergence theorem.

7.3. Stochastic Stability: Examples and Analysis. Theorem 7.7 describes the rate of decay of the stationary distribution weights as η approaches 0 and N approaches infinity. If the main concern is with the states that are likely to be observed with some frequency over the infinite horizon, then one can focus on states $\chi \in [0,1]$ with $\Delta I(\chi) = 0$, since only neighborhoods of such states receive nonnegligible mass in $\mu^{N,\eta}$ for large N in small η. We therefore call state χ *weakly stochastically stable* if it maximizes the ordinal potential I on the unit interval, and we call state χ *uniquely stochastically stable* if it is the unique maximizer of I on the unit interval. We now investigate in greater detail how a game's payoff function and the revision protocol's cost function interact to determine the stochastically stable states.

Stochastic stability analysis is most interesting when it allows us to select among multiple strict equilibria. For this reason, we focus the analysis to come on coordination games. The two-strategy population game $F : [0,1] \to \mathbb{R}^2$ is a *coordination game* if there is a state $\chi^* \in (0,1)$ such that

$$\operatorname{sgn}(F_\Delta(\chi)) = \operatorname{sgn}(\chi - \chi^*) \quad \text{for all } \chi \neq \chi^*.$$

Any ordinal potential function I for a coordination game is quasiconvex, with local maximizers at each boundary state. Because $I(0) \equiv 0$ by definition, Theorem 7.7 implies the following result.

COROLLARY 7.8. *Suppose that the limit game F is a coordination game. Then state 1 is uniquely stochastically stable in both double limits if $I(1) > 0$, while state 0 is uniquely stochastically stable in both double limits if $I(1) < 0$.*

The next two examples, which revisit two games introduced in the previous chapter, show that the identity of the stochastically stable state may or may not depend on the revision protocol the agents employ.

Example 7.9. Stag Hunt revisited. In Example 5.4, we considered stochastic evolution in the Stag Hunt game

$$A = \begin{pmatrix} h & h \\ 0 & s \end{pmatrix},$$

where $s > h > 0$. When a continuous population of agents are matched to play this game, their expected payoffs are given by $F_H(\chi) = h$ and $F_S(\chi) = s\chi$, where χ denotes the proportion of agents playing Stag. This coordination game has two pure Nash equilibria, as well as a mixed Nash equilibrium that puts weight $\chi^* = \frac{h}{s}$ on Stag.

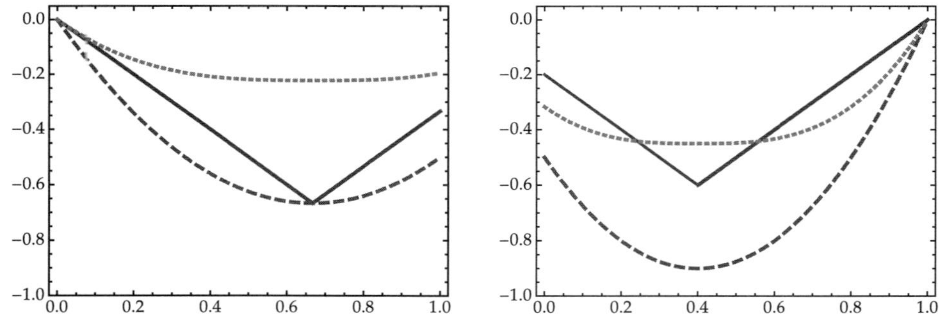

(I) $h = 2,\ s = 3$ (II) $h = 2,\ s = 5$

FIGURE 5. The ordinal potentials ΔI_{sgn} (solid), ΔI_1 (dashed), and ΔI_2 (dotted) for Stag Hunt.

The ordinal potentials for the BRM, logit, and probit protocols in this game are

$$\begin{aligned} I_{\text{sgn}}(x) &= |x - x^*| - x^*, \\ I_1(x) &= \tfrac{s}{2}x^2 - hx, \text{ and} \\ I_2(x) &= \begin{cases} -\frac{s^2}{12}x^3 + \frac{hs}{4}x^2 - \frac{h^2}{4}x & \text{if } x \le x^*, \\ \frac{s^2}{12}x^3 - \frac{hs}{4}x^2 + \frac{h^2}{4}x - \frac{h^3}{6s} & \text{if } x > x^*. \end{cases} \end{aligned}$$

Figure 5 presents the normalized functions ΔI_{sgn}, ΔI_1, and ΔI_2 for two specifications of payoffs: $h = 2$ and $s = 3$ (in (i)), and $h = 2$ and $s = 5$ (in (ii)). For any choices of $s > h > 0$, ΔI is symmetric about its minimizer, the mixed Nash equilibrium $x^* = \frac{h}{s}$. As a result, the three protocols always agree about equilibrium selection: the all-Hare equilibrium is uniquely stochastically stable when $x^* > \frac{1}{2}$ (or, equivalently, when $2h > s$), while the all-Stag equilibrium is uniquely stochastically stable when the reverse inequality holds. ♦

Example 7.10. Nonlinear Stag Hunt revisited. In Example 5.5, we introduced the nonlinear Stag Hunt game with payoff functions $F_H(x) = h$ and $F_S(x) = sx^2$, with x again representing the proportion of agents playing Stag. This game has two pure Nash equilibria and a mixed equilibrium at $x^* = \sqrt{h/s}$. The payoffs and mixed equilibria for $h = 2$ and various choices of s are graphed in Figure 6.

The ordinal potentials for the BRM, logit, and probit models are given by

$$\begin{aligned} I_{\text{sgn}}(x) &= |x - x^*| - x^*, \\ I_1(x) &= \tfrac{s}{3}x^3 - hx, \text{ and} \\ I_2(x) &= \begin{cases} -\frac{s^2}{20}x^5 + \frac{hs}{6}x^3 - \frac{h^2}{4}x & \text{if } x \le x^*, \\ \frac{s^2}{20}x^5 - \frac{hs}{6}x^3 + \frac{h^2}{4}x - \frac{4h^2x^*}{15} & \text{if } x > x^*. \end{cases} \end{aligned}$$

Figure 7 presents the functions ΔI_{sgn}, ΔI_1, and ΔI_2 for $h = 2$ and for various choices of s.

When s is at its lowest level of 5, coordination on Stag is at its least appealing. Since $x^* = \sqrt{2/5} \approx .6325$, the basin of attraction of the all-Hare equilibrium is

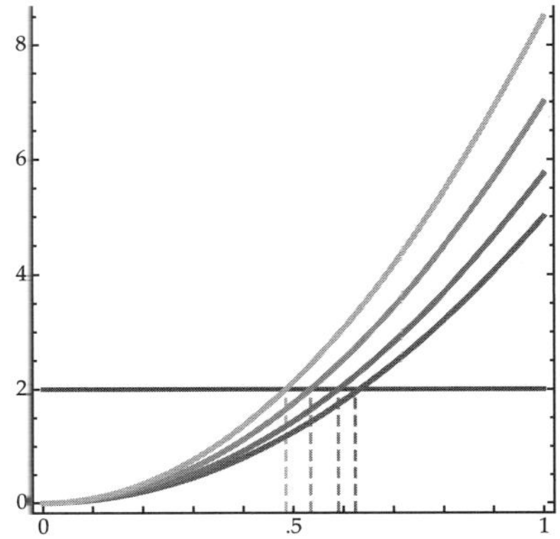

FIGURE 6. Payoffs and mixed equilibria in Nonlinear Stag Hunt when $h = 2$ and $s = 5$, 5.75, 7, and 8.5.

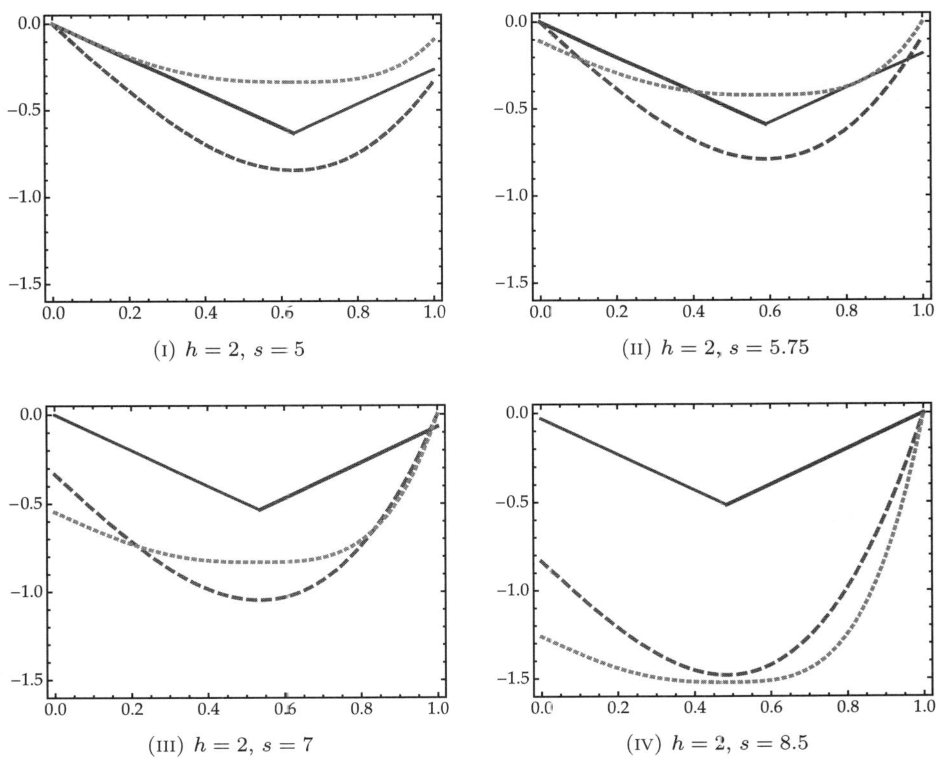

(I) $h = 2$, $s = 5$

(II) $h = 2$, $s = 5.75$

(III) $h = 2$, $s = 7$

(IV) $h = 2$, $s = 8.5$

FIGURE 7. The ordinal potentials ΔI_{sgn} (solid), ΔI_1 (dashed), and ΔI_2 (dotted) for Nonlinear Stag Hunt.

considerably larger than that of the all-Stag equilibrium. Figure 7(i) illustrates that coordination on Hare is stochastically stable under all three protocols.

If we make coordination on Stag somewhat more attractive by increasing s to 5.75, the mixed equilibrium becomes $\chi^* = \sqrt{2/5.75} \approx .5898$. The all-Hare equilibrium remains stochastically stable under the BRM and logit rules, but all-Stag becomes stochastically stable under the probit rule (Figure 7(ii)).

Increasing s further to 7 shifts the mixed equilibrium closer to the midpoint of the unit interval ($\chi^* = \sqrt{2/7} \approx .5345$). The BRM rule continues to select all-Hare, while the probit and logit rules both select all-Stag (Figure 7(iii)).

Finally, when $s = 8.5$, the all-Stag equilibrium has the larger basin of attraction ($\chi^* = \sqrt{2/8.5} \approx .4851$). At this point, coordination on Stag becomes attractive enough that all three protocols select the all-Stag equilibrium (Figure 7(iv)).

Why as we increase the value of s does the transition to selecting all-Stag occur first for the probit rule, then for the logit rule, and finally for the BRM rule? Examining Figure 6, we see that increasing s not only shifts the mixed Nash equilibrium to the left, but also markedly increases the payoff advantage of Stag at states where it is optimal. Since the cost function of the probit rule is the most sensitive to payoff differences, its equilibrium selection changes at the lowest level of s. The next selection to change is that of the (moderately sensitive) logit rule, and the last is the selection of the (insensitive) BRM rule. ♦

7.4. Risk Dominance, Stochastic Dominance, and Stochastic Stability. Building on these examples, we now seek general conditions on payoffs that ensure stochastic stability under all noisy best response protocols.

Example 7.9 showed that in the Stag Hunt game with linear payoffs, the noisy best response rules we considered always selected the equilibrium with the larger basin of attraction. The reason for this is easy to explain. Linearity of payoffs, along with the fact that the relative cost function $\tilde{\kappa}$ is sign-preserving and odd (see equation (26)), implies that the ordinal potential function I is symmetric about the mixed equilibrium χ^*, where it attains its minimum value. If, for example, χ^* is less than $\frac{1}{2}$, so that pure equilibrium 1 has the larger basin of attraction, then $I(1)$ exceeds $I(0)$, implying that state 1 is uniquely stochastically stable. Similarly, if χ^* exceeds $\frac{1}{2}$, then $I(0)$ exceeds $I(1)$, and state 0 is uniquely stochastically stable.

With this motivation, we call strategy i *strictly risk dominant* in the two-strategy coordination game F if the set of states where it is the unique best response is larger than the corresponding set for strategy $j \neq i$. Thus, if F has mixed equilibrium $\chi^* \in (0,1)$, then strategy 0 is strictly risk dominant if $\chi^* > \frac{1}{2}$, and strategy 1 is strictly risk dominant if $\chi^* < \frac{1}{2}$. If the relevant inequality holds weakly in either case, we call the strategy in question *weakly risk dominant*.

The foregoing arguments yield the following result, in which we denote by e_i the state at which all agents play strategy i.

COROLLARY 7.11. *Suppose that the limit game F is a coordination game with linear payoffs. Then*

(*i*) *State e_i is weakly stochastically stable under every noisy best response protocol if and only if strategy i is weakly risk dominant in F.*

(*ii*) *If strategy i is strictly risk dominant in F, then state e_i is uniquely stochastically stable under every noisy best response protocol.*

Example 7.10 shows that once we turn to games with nonlinear payoffs, risk dominance only characterizes stochastic stability under the BRM rule. In any coordination game with mixed equilibrium χ^*, the ordinal potential function for the BRM rule is $I_{\text{sgn}}(\chi) = |\chi - \chi^*| - \chi^*$. This function is minimized at χ^*, and increases at a unit rate as one moves away from χ^* in either direction, reflecting the fact that under the BRM rule, the probability of a suboptimal choice is independent of its payoff consequences. Clearly, whether $I_{\text{sgn}}(1)$ is greater than $I_{\text{sgn}}(0)$ depends only on whether χ^* is less than $\frac{1}{2}$. We therefore have

COROLLARY 7.12. *Suppose that the limit game F is a coordination game and that σ^η is the BRM rule. Then*
- (*i*) *State e_i is weakly stochastically stable if and only if strategy i is weakly risk dominant in F.*
- (*ii*) *If strategy i is strictly risk dominant in F, then state e_i is uniquely stochastically stable.*

Once one moves beyond the BRM rule and linear payoffs, risk dominance is no longer a necessary or sufficient condition for stochastic stability. In what follows, we introduce a natural refinement of risk dominance that serves this role.

To work toward our new definition, let us first observe that any function on the unit interval $[0, 1]$ can be viewed as a random variable by regarding the interval as a sample space endowed with Lebesgue measure λ. With this interpretation in mind, we define the *advantage distribution* of strategy i to be the cumulative distribution function of the payoff advantage of strategy i over the alternative strategy $j \neq i$:

$$G_i(a) = \lambda(\{\chi \in [0,1] : F_i(\chi) - F_j(\chi) \leq a\}).$$

We let $\bar{G}_i$ denote the corresponding decumulative distribution function:

$$\bar{G}_i(a) = \lambda(\{\chi \in [0,1] : F_i(\chi) - F_j(\chi) > a\}) = 1 - G_i(a).$$

In words, $\bar{G}_i(a)$ is the measure of the set of states at which the payoff to strategy i exceeds the payoff to strategy j by more than a.

It is easy to restate the definition of risk dominance in terms of the advantage distribution.

OBSERVATION 7.13. *Let F be a coordination game. Then strategy i is weakly risk dominant if and only if $\bar{G}_i(0) \geq \bar{G}_j(0)$, and strategy i is strictly risk dominant if and only if $\bar{G}_i(0) > \bar{G}_j(0)$.*

To obtain our refinement of risk dominance, we require not only that strategy i be optimal at a larger set of states than strategy j, but also that strategy i have a payoff advantage of at least a at a larger set of states than strategy j for every $a \geq 0$. More precisely, we say that strategy i is *weakly stochastically dominant* in the coordination game F if $\bar{G}_i(a) \geq \bar{G}_j(a)$ for all $a \geq 0$. If in addition $\bar{G}_i(0) > \bar{G}_j(0)$, we say that strategy i is *strictly stochastically dominant*. The notion of stochastic dominance for strategies proposed here is obtained by applying the usual definition of stochastic dominance from utility theory (see Border (2001)) to the strategies' advantage distributions.

Theorem 7.14 shows that stochastic dominance is both sufficient and necessary to ensure stochastic stability under every noisy best response rule.

THEOREM 7.14. *Suppose that the limit game F is a coordination game. Then* (*i*) *State e_i is weakly stochastically stable under every noisy best response protocol if and only if strategy i is weakly stochastically dominant in F.*

(*ii*) *If strategy i is strictly stochastically dominant in F, then state e_i is uniquely stochastically stable under every noisy best response protocol.*

The idea behind Theorem 7.14 is simple. The definitions of I, $\tilde{\kappa}$, κ, F_Δ, and G_i imply that

$$
\begin{aligned}
I(1) &= \int_0^1 \tilde{\kappa}(F_\Delta(y))\,\mathrm{d}y \\
&= \int_0^1 \kappa(F_1(y) - F_0(y))\,\mathrm{d}y - \int_0^1 \kappa(F_0(y) - F_1(y))\,\mathrm{d}y \\
&= \int_{-\infty}^{\infty} \kappa(a)\,\mathrm{d}G_1(a) - \int_{-\infty}^{\infty} \kappa(a)\,\mathrm{d}G_0(a).
\end{aligned} \tag{29}
$$

As we have seen, whether state e_1 or state e_0 is stochastically stable depends on whether $I(1)$ is greater than or less than $I(0) = 0$. This in turn depends on whether the value of the first integral in the final line of (29) exceeds the value of the second integral. Once we recall that the cost function κ is monotone, Theorem 7.14 reduces to a variation on the standard characterization of first-order stochastic dominance: namely, that distribution G_1 stochastically dominates distribution G_0 if and only if $\int \kappa\,\mathrm{d}G_1 \geq \int \kappa\,\mathrm{d}G_0$ for every nondecreasing function κ.

8. Further Developments

The analyses in the previous sections have focused on evolution in two-strategy games, mostly under noisy best response protocols. Two-strategy games have the great advantage of generating birth-and-death processes. Because such processes are reversible, their stationary distributions can be computed explicitly, greatly simplifying the analysis. Other work in stochastic evolutionary game theory focusing on birth-and-death chain models includes Binmore and Samuelson (1997), Maruta (2002), Blume (2003), and Sandholm (2011). The only many-strategy evolutionary game environments known to generate reversible processes are potential games (Monderer and Shapley (1996); Sandholm (2001)), with agents using either the standard (Example 5.2) or imitative versions of the logit choice rule; see Blume (1993, 1997) and Sandholm (2011) for analyses of these models.

Once one moves beyond reversible settings, obtaining exact formulas for the stationary distribution is generally impossible, and one must attempt to determine the stochastically stable states by other means. In general, the available techniques for doing so are descendants of the analyses of sample path large deviations due to Freidlin and Wentzell (1998), and introduced to evolutionary game theory by Kandori et al. (1993) and Young (1993).

One portion of the literature considers *small noise limits*, determining which states retain mass in the stationary distribution as the amount of noise in agents' decisions vanishes. The advantage of this approach is that the set of population states stays fixed and finite. This makes it possible to use the ideas of Freidlin and Wentzell (1998) with few technical complications, but also without the computational advantages that a continuous state space can provide. Many of the analyses of small noise limits focus on the best response with mutations model (Example 5.1); see Kandori et al. (1993), Young (1993, 1998), Kandori and Rob

(1995, 1998), Ellison (2000), and Beggs (2005). Analyses of other important models include Myatt and Wallace (2003), Fudenberg and Imhof (2006, 2008), Dokumacı and Sandholm (2011), and Staudigl (2011).

Alternatively, one can consider *large population limits*, examining the behavior of the stationary distribution as the population size approaches infinity. Here, as one increases the population size, the set of population states becomes an increasingly fine grid in the simplex X. While this introduces some technical challenges, it also allows one to use methods from optimal control theory in the analysis of sample path large deviations. The use of large population limits in stochastic evolutionary models was first proposed by Binmore and Samuelson (1997) and Blume (2003) in two-strategy settings. Analyses set in more general environments include Benaïm and Weibull (2003) and Benaïm and Sandholm (2011), both of which build on results in Benaïm (1998). The analysis of infinite-horizon behavior in the large population limit is still at an early stage of development, and so offers a promising avenue for future research.

References

Beckmann, M., McGuire, C. B., and Winsten, C. B. (1956). *Studies in the Economics of Transportation*. Yale University Press, New Haven.

Beggs, A. W. (2005). On the convergence of reinforcement learning. *Journal of Economic Theory*, 122:1–36.

Benaïm, M. (1998). Recursive algorithms, urn processes, and the chaining number of chain recurrent sets. *Ergodic Theory and Dynamical Systems*, 18:53–87.

Benaïm, M. and Sandholm, W. H. (2011). Large deviations, reversibility, and equilibrium selection under evolutionary game dynamics. Unpublished manuscript, Université de Neuchâtel and University of Wisconsin.

Benaïm, M. and Weibull, J. W. (2003). Deterministic approximation of stochastic evolution in games. *Econometrica*, 71:873–903.

Binmore, K. and Samuelson, L. (1997). Muddling through: Noisy equilibrium selection. *Journal of Economic Theory*, 74:235–265.

Björnerstedt, J. and Weibull, J. W. (1996). Nash equilibrium and evolution by imitation. In Arrow, K. J. et al., editors, *The Rational Foundations of Economic Behavior*, pages 155–181. St. Martin's Press, New York.

Blume, L. E. (1993). The statistical mechanics of strategic interaction. *Games and Economic Behavior*, 5:387–424.

Blume, L. E. (1997). Population games. In Arthur, W. B., Durlauf, S. N., and Lane, D. A., editors, *The Economy as an Evolving Complex System II*, pages 425–460. Addison-Wesley, Reading, MA.

Blume, L. E. (2003). How noise matters. *Games and Economic Behavior*, 44:251–271.

Border, K. C. (2001). Comparing probability distributions. Unpublished manuscript, Caltech.

Brown, G. W. and von Neumann, J. (1950). Solutions of games by differential equations. In Kuhn, H. W. and Tucker, A. W., editors, *Contributions to the Theory of Games I*, volume 24 of *Annals of Mathematics Studies*, pages 73–79. Princeton University Press, Princeton.

Dokumacı E. and Sandholm, W. H. (2011). Large deviations and multinomial probit choice. *Journal of Economic Theory*, forthcoming.

Ellison G. (2000). Basins of attraction, long run equilibria, and the speed of step-by-step evolution. *Review of Economic Studies*, 67:17–45.

Freidlin, M. I. and Wentzell, A. D. (1998). *Random Perturbations of Dynamical Systems.* Springer, New York, second edition.

Fudenberg, D. and Imhof, L. A. (2006). Imitation processes with small mutations. *Journal of Economic Theory*, 131:251–262.

Fudenberg, D. and Imhof, L. A. (2008). Monotone imitation dynamics in large populations. *Journal of Economic Theory*, 140:229–245.

Fudenberg, D. and Levine, D. K. (1998). *The Theory of Learning in Games.* MIT Press, Cambridge.

Gilboa, I. and Matsui, A. (1991). Social stability and equilibrium. *Econometrica*, 59:859–867.

Helbing, D. (1992). A mathematical model for behavioral changes by pair interactions. In Haag, G., Mueller, U., and Troitzsch, K. G., editors, *Economic Evolution and Demographic Change: Formal Models in Social Sciences*, pages 330–348. Springer, Berlin.

Hofbauer, J. (1995). Imitation dynamics for games. Unpublished manuscript, University of Vienna.

Hofbauer, J. and Sandholm, W. H. (2002). On the global convergence of stochastic fictitious play. *Econometrica*, 70:2265–2294.

Hofbauer, J. and Sigmund, K. (1988). *Theory of Evolution and Dynamical Systems.* Cambridge University Press, Cambridge.

Hofbauer, J. and Sigmund, K. (1998). *Evolutionary Games and Population Dynamics.* Cambridge University Press, Cambridge.

Hofbauer, J. and Sigmund, K. (2003). Evolutionary game dynamics. *Bulletin of the American Mathematical Society (New Series)*, 40:479–519.

Kandori, M., Mailath, G. J., and Rob, R. (1993). Learning, mutation, and long run equilibria in games. *Econometrica*, 61:29–56.

Kandori, M. and Rob, R. (1995). Evolution of equilibria in the long run: A general theory and applications. *Journal of Economic Theory*, 65:383–414.

Kandori, M. and Rob, R. (1998). Bandwagon effects and long run technology choice. *Games and Economic Behavior*, 22:84–120.

Kurtz, T. G. (1970). Solutions of ordinary differential equations as limits of pure jump Markov processes. *Journal of Applied Probability*, 7:49–58.

Maruta, T. (2002). Binary games with state dependent stochastic choice. *Journal of Economic Theory*, 103:351–376.

Maynard Smith, J. (1982). *Evolution and the Theory of Games.* Cambridge University Press, Cambridge.

Monderer, D. and Shapley, L. S. (1996). Potential games. *Games and Economic Behavior*, 14:124–143.

Moran, P. A. P. (1962). *The Statistical Processes of Evolutionary Theory.* Clarendon Press, Oxford.

Myatt, D. P. and Wallace, C. C. (2003). A multinomial probit model of stochastic evolution. *Journal of Economic Theory*, 113:286–301.

Norris, J. R. (1997). *Markov Chains.* Cambridge University Press, Cambridge.

Nowak, M. A. (2006). *Evolutionary Dynamics: Exploring the Equations of Life.* Belknap/Harvard, Cambridge.

Nowak, M. A., Sasaki, A., Taylor, C., and Fudenberg, D. (2004). Emergence of cooperation and evolutionary stability in finite populations. *Nature*, 428:646–650.

Sandholm, W. H. (2001). Potential games with continuous player sets. *Journal of Economic Theory*, 97:81–108.

Sandholm, W. H. (2003). Evolution and equilibrium under inexact information. *Games and Economic Behavior*, 44:343–378.

Sandholm, W. H. (2007). Simple formulas for stationary distributions and stochastically stable states. *Games and Economic Behavior*, 59:154–162.

Sandholm, W. H. (2009). Evolutionary game theory. In Meyers, R. A., editor, *Encyclopedia of Complexity and Systems Science*, pages 3176–3205. Springer, Heidelberg.

Sandholm, W. H. (2010a). Orders of limits for stationary distributions, stochastic dominance, and stochastic stability. *Theoretical Economics*, 5:1–26.

Sandholm, W. H. (2010b). Pairwise comparison dynamics and evolutionary foundations for Nash equilibrium. *Games*, 1:3–17.

Sandholm, W. H. (2010c). *Population Games and Evolutionary Dynamics.* MIT Press, Cambridge.

Sandholm, W. H. (2011). Stochastic imitative game dynamics with committed agents. Unpublished manuscript, University of Wisconsin.

Schlag, K. H. (1998). Why imitate, and if so, how? A boundedly rational approach to multi-armed bandits. *Journal of Economic Theory*, 78:130–156.

Smith, M. J. (1984). The stability of a dynamic model of traffic assignment—an application of a method of Lyapunov. *Transportation Science*, 18:245–252.

Staudigl, M. (2011). Stochastic stability in binary choice coordination games. Unpublished manuscript, European University Institute.

Taylor, P. D. and Jonker, L. (1978). Evolutionarily stable strategies and game dynamics. *Mathematical Biosciences*, 40:145–156.

Traulsen, A. and Hauert, C. (2009). Stochastic evolutionary game dynamics. In Schuster, H. G., editor, *Reviews of Nonlinear Dynamics and Complexity*, volume 2, pages 25–61. Wiley, New York.

Weibull, J. W. (1995). *Evolutionary Game Theory.* MIT Press, Cambridge.

Young, H. P. (1993). The evolution of conventions. *Econometrica*, 61:57–84.

Young, H. P. (1998). *Individual Strategy and Social Structure.* Princeton University Press, Princeton.

Department of Economics, University of Wisconsin, 1180 Observatory Drive, Madison, WI 53706, USA.

E-mail address: whs@ssc.wisc.edu

Proceedings of Symposia in Applied Mathematics
Volume **69**, 2011

Evolution of Cooperation in Finite Populations

Sabin Lessard

ABSTRACT. The Iterated Prisoner's Dilemma with an additive effect on viability selection as payoff is used to study the evolution of cooperation in finite populations. A condition for weak selection to favor Tit-for-Tat replacing Always-Defect when introduced as a single mutant strategy in a well-mixed population is deduced from the sum of all future expected changes in frequency. It is shown by resorting to coalescent theory that the condition reduces to the one-third law of evolution in the realm of the Kingman coalescent in the limit of a large population size. The condition proves to be more stringent when the reproductive success of an individual is a random variable having a highly skewed probability distribution. An explanation of the one-third law of evolution based on the notion of projected average excess in payoff is provided. A two-timescale argument is applied for group-structured populations. The condition is found to be less stringent in the case of uniform dispersal of offspring followed by interactions within groups. The condition becomes even less stringent if dispersal occurs after interactions so that there are differential contributions of groups in offspring. On the other hand, the condition is strengthened by a highly skewed probability distribution for the contribution of a group in offspring.

1. Introduction

Although cooperation is widespread in nature, its evolution is difficult to explain. The main problem is that cooperation did not always exist and before being common in a population it must have been rare. But the advantage of cooperation when rare is not obvious. In order to study the advantage of cooperation and understand its evolution, we will consider a game-theoretic framework based on pairwise interactions.

In the Prisoner's Dilemma (PD) two accomplices in committing a crime are arrested and each one can either defect (D) by testifying against the other or cooperate with the other (C) by remaining silent. Each of the accomplices receives a light sentence corresponding to some reward (R) when both cooperate, compared to a heavy sentence corresponding to a punishment (P) when both defect. When one defects and the other cooperates the defector receives a lighter sentence represented by some temptation (T), while the cooperator receives a heavier sentence represented by the sucker's payoff (S). Therefore, the payoffs in the PD game satisfy the inequalities $T > R > P > S$. The situation is summarized in Fig. 1 with some particular values for the different payoffs.

Note that strategy D is the best reply to itself, since the payoff to D against D exceeds the payoff to C against D. Actually the payoff to C is smaller than the payoff to D whatever

2000 *Mathematics Subject Classification.* Primary 60C05; Secondary 92D15 .
Research supported in part by NSERC Grant 8833.

*C*ooperate	*R*eward 5	*S*ucker's payoff 1
*D*efect	*T*emptation 14	*P*unishment 3
against	*C*ooperate	*D*efect

FIGURE 1. Payoffs in the PD game with some particular values.

the strategy of the opponent is. If pairwise interactions occur at random in an infinite population, then the expected payoff to C can only be smaller than the expected payoff to D. Moreover, if the reproductive success of an individual is an increasing function of the payoff and true breeding is assumed so that an offspring uses the same strategy as its parent, then C is not expected to increase in frequency.

In order to find conditions that could favor the evolution of cooperation the PD game is extended by assuming n rounds of the game between the same players. This is known as the Iterated Prisoner's Dilemma (IPD). Then two sequential strategies are considered: Tit-for-Tat, represented by A, and Always-Defect, represented by B. Always-Defect consists obviously in defecting in every round, while Tit-for-Tat consists in cooperating in the first round and then using the previous move of the opponent in the next rounds. Note that two players using Tit-for-Tat will always cooperate. Moreover, Tit-for-Tat has proved to do better than any other sequential strategy in computer experiments. See, e.g., Axelrod (1984), Hofbauer and Sigmund (1998, Chap. 9), McNamara *et al.* (2004), and references therein for more details, variants and historical perspectives.

Let us assume that the payoffs in the different repetitions of the IPD game are additive. Then the payoffs to A against A, A against B, B against A, and B against B, denoted by a, b, c, and d, respectively, take the expressions given in Fig. 2. Note that these payoffs satisfy the inequalities $a > c > d > b$ as soon as the number of repetitions is large enough, that is,

$$n > \frac{T-P}{R-P}. \tag{1.1}$$

This condition guarantees that A is the best reply to itself, since then $a > c$ which means that the payoff to A against A exceeds the payoff to B against A. Similarly the inequality $d > b$ means that B is the best reply to itself. This is the situation, for instance, when $n = 10$ with the payoffs of the PD game given in Fig. 1. The consequence of this is that the expected payoff to A will exceed the expected payoff to B in an infinite population with random pairwise interactions if the frequency of A exceeds some threshold value between 0 and 1.

Tit-for-Tat (A)	$a = Rn$ 50	$b = S + P(n-1)$ 28
Always-Defect (B)	$c = T + P(n-1)$ 41	$d = Pn$ 30
against	A	B

FIGURE 2. Payoffs in the IPD game with particular values in the case $n = 10$ with the numerical payoffs of the PD game given in Fig. 1.

As a matter of fact, if the frequencies of A and B in an infinite population are x and $1-x$, respectively, then the expected payoffs to A and B are

$$w_A(x) = ax + b(1-x) \tag{1.2}$$

and

$$w_B(x) = cx + d(1-x), \tag{1.3}$$

respectively. Therefore, $w_A(x) > w_B(x)$ if and only if

$$x > \frac{d-b}{a-b-c+d} = x^*. \tag{1.4}$$

With the expressions of the different payoffs given in Fig. 2, we find that

$$x^* = \frac{P-S}{(P-S)+(R-P)\left(n - \frac{T-P}{R-P}\right)}. \tag{1.5}$$

This threshold value for x decreases from 1 to 0 as n increases from $(T-P)/(R-P)$ to infinity, but remains always positive. This suggests that the frequency of A in an infinite population can increase, but only if the initial frequency is high enough.

2. Dynamics in an infinite population

Consider an infinite haploid population undergoing discrete, non-overlapping generations and suppose random pairwise interactions among the offspring of the same generation. These interactions are assumed to have an additive effect on viability. More precisely the probability for an individual to survive from conception to maturity, and then to contribute to the next generation, is proportional to some fitness given in the form

$$\text{fitness} = 1 + s \times \text{payoff}. \tag{2.1}$$

Here, 1 is an arbitrary reference value and $s \geq 0$ represents an intensity of viability selection whose coefficient is the payoff to the individual. The intensity of selection will be assumed small throughout the paper. The case $s = 0$ corresponds to neutrality.

Let $x(t)$ be the frequency of A in generation t before selection. As a result of random pairwise interactions, the probability for an individual of type A to survive will be $1+sw_A(x(t))$ compared to $1+sw_B(x(t))$ for an individual of type B. Then the frequency of A in generation t after selection will be

$$\tilde{x}(t) = \frac{x(t)(1+sw_A(x(t)))}{1+sw(x(t))}, \tag{2.2}$$

where

$$w(x(t)) = x(t)w_A(x(t)) + (1-x(t))w_B(x(t)) \tag{2.3}$$

is the mean payoff in generation t. After reproduction and in the absence of mutation, this frequency will be also the frequency of A in the offspring of generation $t+1$, that is, $x(t+1) = \tilde{x}(t)$. Therefore, the change in the frequency of A before selection from generation t to generation $t+1$, represented by $\Delta x(t) = x(t+1) - x(t)$, will be given by

$$\Delta x(t)\tilde{x}(t) - x(t) = \frac{sx(t)(1-x(t))(w_A(x(t)) - w_B(x(t)))}{1+sw(x(t))}, \tag{2.4}$$

where

$$w_A(x(t)) - w_B(x(t)) = (a-b-c+d)(x(t)-x^*). \tag{2.5}$$

We conclude that $\Delta x(t) = 0$ if and only if $x(t) = 0, 1$ or x^*, which are the stationary states. Moreover, since $a-b-c+d > 0$ and $0 < x^* < 1$, we have that $\Delta x(t) > 0$ if $x(t) > x^*$, while $\Delta x(t) < 0$ if $x(t) < x^*$. Therefore, $x(t)$ increases as $t \to \infty$ if $x(0) > x^*$, while it decreases if $x(0) < x^*$. Actually $x(t)$ increases to 1 in the former case, while $x(t)$ decreases to 0 in the latter case, since the limit of $x(t)$ as $t \to \infty$ must be a stationary state by continuity.

Let us summarize.

Proposition 1. *Consider the IPD game in Fig. 2 with a number of rounds satisfying $n > (T-P)/(R-P)$ so that the payoffs to A and B satisfy $a > c$ and $d > b$, or more generally any two-player two-strategy matrix game with strategies A and B being the best replies to themselves. Assume random pairwise interactions in an infinite population undergoing discrete non-overlapping generations and viability selection of intensity s with coefficient given by the payoff. The frequency of A in generation t before selection, $x(t)$, satisfies*

$$x(t) \uparrow 1 \textit{ if } x(0) > x^\star \textit{ and } x(t) \downarrow 0 \textit{ if } x(0) < x^\star, \tag{2.6}$$

where

$$x^* = \frac{d-b}{a-b-c+d} \tag{2.7}$$

is a stationary state in $(0,1)$ for the deterministic dynamics.

Proposition 1 means that x^* is an unstable polymorphic equilibrium, while 0 and 1 are monomorphic stable equilibria. Unfortunately this cannot explain the spread of A from an initial low frequency following its introduction as a rare mutant strategy.

3. Fixation probability in a finite population

In a finite population, random drift that results from sampling effects can ultimately bring the frequency of A to fixation from any low initial frequency. In this section we consider the probability of this event.

Each generation starts with N parents labeled from 1 to N. These produce virtually infinite numbers of offspring identical to themselves in the relative proportions $\pi_1, \ldots, \pi_N$, respectively. The population size N is assumed to be finite and constant. The proportions $\pi_1, \ldots, \pi_N$ are exchangeable random variables. This means that the joint distribution is invariant under any permutation. Furthermore, they satisfy $0 \leq \pi_i \leq 1$ for $i = 1, \ldots, N$ and $\sum_{i=1}^N \pi_i = 1$. In particular this implies that the expected proportion of offspring produced by each parent is the same. It is given by

$$E(\pi_1) = N^{-1} \sum_{i=1}^{N} E(\pi_i) = N^{-1} E\left(\sum_{i=1}^{N} \pi_i\right) = N^{-1}. \tag{3.1}$$

Moreover, it is assumed that

$$c_N = NE(\pi_1^2) = \sum_{i=1}^{N} E(\pi_i^2) \to 0 \tag{3.2}$$

as $N \to \infty$. This says that the probability for two offspring chosen at random without replacement to have the same parent tends to 0 as the population size increases.

The Wright-Fisher model (Fisher 1930, Wright 1931) corresponds to the situation where $\pi_i = N^{-1}$ for $i = 1, \ldots, N$. In this case, we have $c_N = N^{-1}$.

A modified Wright-Fisher model with a skewed distribution of progeny size can be obtained by allowing for $\pi_i = \psi$ for some i chosen at random and $\pi_j = (1-\psi)(N-1)^{-1}$ for every $j \neq i$, for some fixed $0 < \psi < 1$. A combination of this reproduction scheme with probability $N^{-\alpha}$ and the Wright-Fisher scheme with the complementary probability $1 - N^{-\alpha}$, and this for each generation, has been considered and applied to oysters for instance (Eldon and Wakeley 2006). In this case, we find that

$$c_N = \frac{1}{N}\left(1 - \frac{1}{N^\alpha}\right) + \frac{1}{N^\alpha}\left(\psi^2 + \frac{(1-\psi)^2}{N-1}\right), \tag{3.3}$$

whose leading term is $\psi^2 N^{-\alpha}$ if $0 < \alpha < 1$, but $(1+\psi)^2 N^{-1}$ if $\alpha = 1$, and N^{-1} if $\alpha > 1$.

The general situation of exchangeable proportions of offspring produced by the N parents each generation corresponds to the Cannings model (Cannings 1974).

The frequency of A in the parents of generation t is represented by a random variable $z(t)$. This random variable can take only the values i/N for $i = 0, 1, \ldots, N$. The frequency of A in the offspring of generation t is represented by $x(t)$, which has the same expected value as $z(t)$. This frequency becomes $\tilde{x}(t)$ as defined in the previous section in the adults of generation t after viability selection as a result of random pairwise interactions among the offspring. Then N adults are chosen at random to be the parents of the offspring of generation $t+1$. The frequency of A in these parents is $z(t+1)$. The conditional distribution of $z(t+1)$ given $x(t)$ is the distribution of a binomial random variable of parameters N and $\tilde{x}(t)$, divided by N. In particular, the conditional expected value of $z(t+1)$ is $\tilde{x}(t)$, which is the same as the conditional expected value of $x(t+1)$. (See Fig. 3 for a schematic representation of the life cycle.)

Actually $z(t)$ for $t \geq 0$ is a Markov chain on the finite state space $\{i/N : i = 0, 1, \ldots, N\}$ with fixation states 0 and 1, all other states being transient. From any initial state $z(0)$, the chain will hit 0 or 1 in a finite time with probability 1 owing to the ergodic theorem. Actually as $t \to \infty$ the chain $z(t)$ will converge in probability to a random variable $z(\infty)$ that takes the value 1 with some probability $u(s)$, which is the probability for the chain to hit 1 before 0, and the value 0 with the complementary probability $1 - u(s)$. Here, $u(s)$ represents the probability of ultimate fixation of A as a function of the intensity of selection.

Note that

$$u(s) = E_s[z(\infty)], \tag{3.4}$$

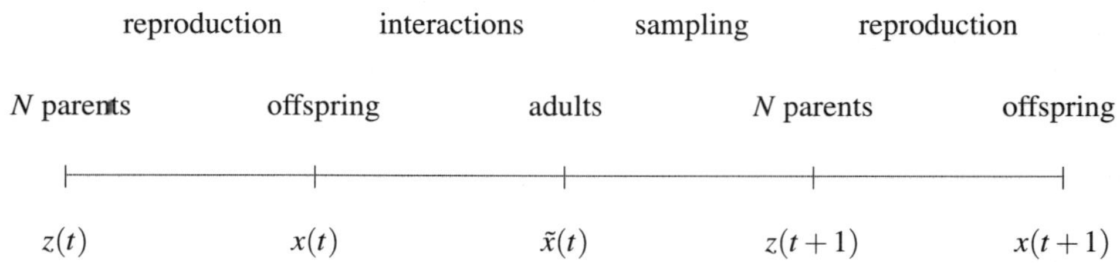

FIGURE 3. Life cycle from generation t to generation $t+1$ and notation for the frequency of A in the whole population at each step.

where E_s denotes expectation as a function of s. Moreover $u(0) = z(0)$, since one of the offspring in the initial generation will be the ancestor of the whole population in the long run, and it will be one offspring chosen at random in the initial generation by symmetry if no selection takes place.

Being uniformly bounded by 1, the chain will also converge in mean. Therefore, we have

$$(3.5) \qquad \begin{aligned} E_s[z(\infty)] &= \lim_{T\to\infty} E_s[z(T))] \\ &= \lim_{T\to\infty} E_s\left[z(0) + \sum_{t=0}^{T}(z(t+1) - z(t))\right] \\ &= z(0) + \lim_{T\to\infty}\sum_{t=0}^{T} E_s[z(t+1) - z(t)] \\ &= z(0) + \sum_{t=0}^{\infty} E_s[z(t+1) - z(t)]. \end{aligned}$$

On the other hand, the tower property of conditional expectation and (2.4) yield

$$(3.6) \qquad \begin{aligned} E_s[z(t+1) - z(t)] &= E_s[x(t+1) - x(t)] \\ &= E_s\left[E_s[x(t+1) - x(t)|x(t)]\right] \\ &= E_s\left[\tilde{x}(t) - x(t)\right] \\ &= s(a-b-c+d)E_s\left[\frac{x(t)(1-x(t))(x(t)-x^*)}{1+sw(x(t))}\right] \\ &= s(a-b-c+d)E\left[x(t)(1-x(t))(x(t)-x^*)\right] + o(s), \end{aligned}$$

where E denotes expectation in the absence of selection, that is, E_s when $s = 0$, while $|o(s)|/s \to 0$ as $s \to 0$. This leads to the approximation

$$(3.7) \qquad u(s) = u(0) + s(a-b-c+d)\sum_{t=0}^{\infty} E\left[x(t)(1-x(t))(x(t)-x^*)\right] + o(s)$$

for the probability of ultimate fixation of A under weak selection.

The above approach was suggested in Rousset (2003) and ascertained in Lessard and Ladret (2007) under mild regularity conditions on the transition probabilities of the Markov chain. Actually it suffices that these probabilities and their derivatives are continuous functions of s at $s = 0$, which is the case here.

The following definition was introduced in Nowak *et al.* (2004).

Definition 1. *Selection favors A replacing B if the probability of ultimate fixation of A is larger in the presence of selection than in the absence of selection.*

The inequality $u(s) > u(0)$ for $s > 0$ small enough guarantees that weak selection favors A replacing B. This will be the case if $u'(0) > 0$, where

$$u'(0) = (a-b-c+d)\sum_{t=0}^{\infty} E\left[x(t)(1-x(t))(x(t)-x^*)\right] \tag{3.8}$$

is the derivative of the fixation probability with respect to the intensity of selection evaluated at $s = 0$. The condition $a-b-c+d > 0$ leads to the following conclusion.

Proposition 2. *Assume that the offspring of generation t are produced in infinite numbers in exchangeable proportions by a fixed finite number N of adults chosen at random in the previous generation and that they undergo viability selection according to the game of Proposition 1. Weak selection favors A replacing B if*

$$x^* < \frac{\sum_{t\geq 0} E[x(t)^2(1-x(t))]}{\sum_{t\geq 0} E[x(t)(1-x(t))]} = \hat{x}, \tag{3.9}$$

where $x(t)$ is the frequency of A in the offspring of generation t and E denotes expectation under neutrality.

Note that the condition for A to be favored for replacement under weak selection is more stringent if the upper bound $\hat{x}$ defined in Proposition 2, which satisfies $0 < \hat{x} < 1$, is closer to 0.

4. Generalized one-third law of evolution

In this section we calculate the upper bound $\hat{x}$ in Proposition 2. This is done under the assumption that A is initially a single mutant, that is, $u(0) = z(0) = N^{-1}$. Moreover, all calculations are made under neutrality.

First note that $E[x(t)(1-x(t))]$ is the probability for two offspring chosen at random without replacement in generation t to be of types A and B in this order. This is a consequence of the tower property of conditional expectation. As a matter of fact, using the indicator random variable $\xi_i(t) = 1$ if the i-th offspring chosen at random without replacement in generation t is of type A, and 0 otherwise, for $i = 1,2$, we have

$$E\left[x(t)(1-x(t))\right] = E\left[E\left[\xi_1(t)(1-\xi_2(t))|x(t)\right]\right] = E\left[\xi_1(t)(1-\xi_2(t))\right]. \tag{4.1}$$

Going backwards in time from generation t to generation 0, we obtain

$$E\left[\xi_1(t)(1-\xi_2(t))\right] = \frac{p_{22}(t+1)}{N}, \tag{4.2}$$

where $p_{22}(t+1) = p_{22}^{t+1}$ is the probability that two offspring chosen at random without replacement in generation t descend from two distinct ancestral parents in generation 0, and $1/N$ the probability that the ancestral parent of the first offspring is of type A. Then necessarily the ancestral parent of the second offspring will be of type B, since A is represented only once in the initial generation. Here, the quantity p_{22} denotes the probability for two offspring chosen at random without replacement in the same generation to have different parents. Therefore,

$$\sum_{t\geq 0} E[x(t)(1-x(t))] = \frac{p_{22}}{N(1-p_{22})}. \tag{4.3}$$

Similarly,

$$E\left[x(t)^2(1-x(t))\right] = E\left[\xi_1(t)\xi_2(t)(1-\xi_3(t))\right] = \frac{p_{32}(t+1)}{3N}, \tag{4.4}$$

where $p_{32}(t+1)$ represents the probability that three offspring chosen at random without replacement in generation t descend from two distinct ancestral parents in generation 0 and $1/3$ is the conditional probability that it is then the first two offspring that descend from the same ancestral parent (see Fig. 4.). Here,

$$p_{32}(t+1) = \sum_{r=0}^{t} p_{33}^{t-r} p_{32} p_{22}^{r} = p_{32}\frac{p_{33}^{t+1} - p_{22}^{t+1}}{p_{33} - p_{22}}, \tag{4.5}$$

where

$$p_{ij} = \sum_{\substack{a_1+\cdots+a_j=i \\ a_1,\ldots,a_j\geq 1}} E\left(\prod_{r=1}^{j} \pi_r^{a_r}\right) \tag{4.6}$$

represents the probability that i offspring chosen at random without replacement in the same generation have j distinct parents. This leads to

$$\sum_{t\geq 0} E\left[x(t)^2(1-x(t))\right] = \frac{p_{32}}{3N(1-p_{22})(1-p_{33})}. \tag{4.7}$$

Finally we obtain

$$\hat{x} = \frac{p_{32}}{3p_{22}(1-p_{33})} \tag{4.8}$$

for the upper bound in Proposition 2. Note that

$$p_{22} = 1 - c_N \to 1, \tag{4.9}$$

as $N \to \infty$, and

$$p_{32} \leq p_{32} + p_{31} = 1 - p_3, \tag{4.10}$$

which complete the proof of the following statement.

Proposition 3. *In the case of a single initial A, the upper bound $\hat{x}$ in the condition given in Proposition 2 for weak selection to favor A replacing B satisfies*

$$\lim_{N\to\infty} \hat{x} = \lim_{N\to\infty} \frac{p_{32}}{3(1-p_{33})} \leq \frac{1}{3}, \tag{4.11}$$

where p_{ij} is the probability that i offspring chosen at random without replacement in the same generation have j distinct parents.

An equality on the right-hand side of the equation in Proposition 3 gives the weakest condition for A to be favored for replacement under weak selection. This is known as the one-third law of evolution (Nowak *et al.* 2004).

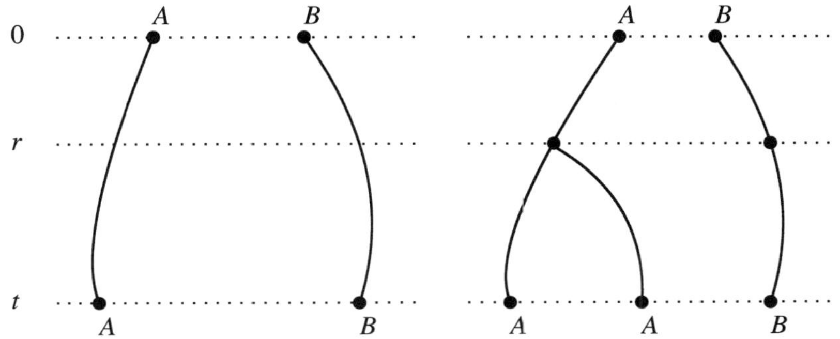

FIGURE 4. Lineages of two offspring of types A,B and three offspring of types A,A,B from generation t to generation 0.

Definition 2. *The one-third law of evolution states that weak selection favors a single A replacing B in the limit of a large population size if* $x^* < 1/3$.

According to Proposition 3, the one-third law of evolution holds if and only if at most two lineages out of three coalesce at a time backwards in time with probability 1 in the limit of a large population size. This is the necessary and sufficient condition for the limiting backward process of the neutral Cannings model with c_N^{-1} generations as unit of time with c_N defined in (3.2) to be the Kingman coalescent (Kingman 1982, Möhle 2000, Möhle and Sagitov 2001).

Let us recall that the number of lineages backwards in time under the Kingman coalescent is a death process on the positive integers with death rate from $k \geq 1$ to $k-1$ given by $\lambda_k = k(k-1)/2$. This means that each pair of lineages coalesces with rate 1 independently of each other.

The above conclusion first drawn in Lessard and Ladret (2007) shows that the one-third law of evolution originally deduced for the Moran model (Nowak *et al.* 2004) and the Wright-Fisher model (Lessard 2005, Imhof and Nowak 2006) holds for a wide class of models. Moreover it shows how the one-third law extends beyond this class. Note that the Moran model (Moran 1958) assumes overlapping generations with one individual replaced at a time, but such models lead to the same conclusion (Lessard and Ladret 2007, Lessard 2007a).

In the case of the Eldon-Wakeley model with probability $N^{-\alpha}$ for a random parent to produce a fraction ψ of all offspring (Eldon and Wakeley 2006), the probability p_{21} that two offspring have the same parent, which is the same as c_N, is given by (3.3), while the probability for three offspring to have the same parent is

$$p_{31} = \frac{1}{N^2}\left(1 - \frac{1}{N^\alpha}\right) + \frac{1}{N^\alpha}\left(\psi^3 + \frac{(1-\psi)^3}{(N-1)^2}\right) \tag{4.12}$$

and the probability for them to have two distinct parents

$$p_{32} = \frac{3}{N}\left(1 - \frac{1}{N}\right)\left(1 - \frac{1}{N^\alpha}\right) + \frac{3(1-\psi)}{N^\alpha}\left(\psi^2 + \frac{1-\psi}{N-1} - \frac{(1-\psi)^2}{(N-1)^2}\right). \tag{4.13}$$

In this case,

$$\lim_{N\to\infty} \frac{p_{32}}{3(1-p_{33})} = \begin{cases} \frac{1}{3} & \text{if } \alpha > 1, \\ \frac{1-\psi}{3-2\psi} & \text{if } \alpha < 1, \\ \frac{1+\psi^2(1-\psi)}{3+\psi^3(3-2\psi)} & \text{if } \alpha = 1. \end{cases} \tag{4.14}$$

The limit is strictly less than 1/3 if and only if $\alpha \leq 1$. This means a more stringent condition for A to be favored for replacement under weak selection when the distribution of progeny size is highly skewed.

Note that $\alpha \leq 1$ is the condition for the limit backward process of the neutral Eldon-Wakeley model with c_N^{-1} generations as unit of time to be a Λ-coalescent allowing for multiple mergers involving more than two lineages (Pitman 1999, Sagitov 1999). In the case $\alpha < 1$, the rate of an m-merger among k lineages is given by

$$\lambda_{k,m} = \binom{k}{m} \psi^{m-2}(1-\psi)^{k-m}, \tag{4.15}$$

for $m = 2, \ldots, k$.

5. Explanation for the one-third law of evolution: projected average excess

The following explanation for the one-third law of evolution in the limit of a large Moran or Wright-Fisher population has been proposed (Ohtsuki *et al.* 2007): in an invasion attempt by a single mutant of type A up to extinction or fixation in the absence of selection, A-players effectively interact on average with B-players twice as often as with A-players. The argument is based on the mean effective sojourn times in the different population states. These can be obtained exactly for the Moran model and approximated for the Wright-Fisher model (Fisher 1930, p. 90).

In this section, we propose another explanation based on the notion of projected average excess (Lessard and Lahaie 2009). This is an extension of the classical notion of average excess in fitness for a gene substitution (Fisher 1930). Here, we consider the excess in payoff for a mutant strategy not only in the current generation but also in all future generations.

First observe that

$$\sum_{t\geq 0} E[x(t)(1-x(t))] = \frac{p_{22}E(S_2)}{N} \tag{5.1}$$

and

$$\sum_{t\geq 0} E\left[x(t)^2(1-x(t))\right] = \frac{E(S_2)-E(S_3)}{2N}, \tag{5.2}$$

where S_2 and S_3 represent the numbers of generations spent backwards in time with two and three lineages, respectively, before the first coalescence event occurs. As a matter of fact, we have

$$E(S_2) = \frac{1}{1-p_{22}} \tag{5.3}$$

and

$$E(S_3) = \frac{1}{1-p_{33}}, \tag{5.4}$$

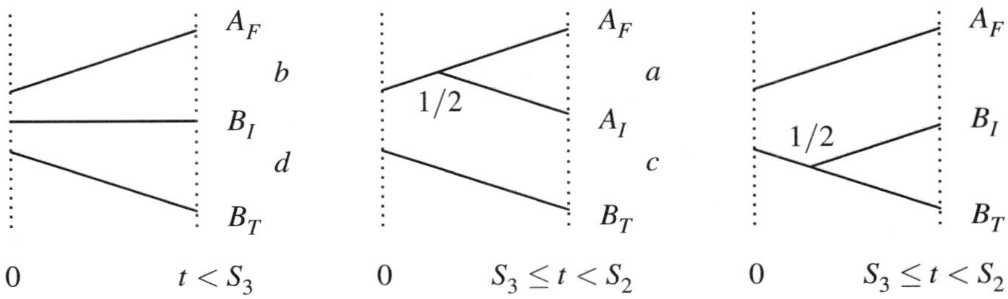

FIGURE 5. Average excess in payoff for A in generation t. The indices F, T and I are used for focal, typical and interacting offspring, respectively. Only typical offspring of type B have to be considered. The coalescence time S_2 is for F and T, while S_3 is for F, T and I.

so that

$$E(S_2) - E(S_3) = \frac{p_{22} - p_{33}}{(1-p_{22})(1-p_{33})}. \tag{5.5}$$

Moreover, we have

$$p_{22} - p_{33} = \frac{2p_{32}}{3}, \tag{5.6}$$

which are two different expressions for the probability that exactly two given offspring out of three chosen at random without replacement have different parents. Therefore, the above equalities agree with the corresponding expressions given in the previous section.

On the other hand, the first derivative of the probability of ultimate fixation of A with respect to the intensity of selection evaluated at $s = 0$ can be written as

$$u'(0) = (a-c)\sum_{t=0}^{\infty} E\left[x(t)^2(1-x(t))\right] + (b-d)\sum_{t=0}^{\infty} E\left[x(t)(1-x(t))^2\right], \tag{5.7}$$

where

$$E\left[x(t)(1-x(t))^2\right] = E\left[x(t)(1-x(t))\right] - E\left[x(t)^2(1-x(t))\right]. \tag{5.8}$$

Then, the above equalities and the assumption that $c_N = 1 - p_{22} \to 0$ as $N \to \infty$ lead to the approximation

$$u'(0) \approx (a-c)\left(\frac{E(S_2)-E(S_3)}{2N}\right) + (b-d)\left(\frac{E(S_2)+E(S_3)}{2N}\right). \tag{5.9}$$

This can be written in the form

$$u'(0) \approx \frac{1}{N}\left\{\left(\frac{a-c+b-d}{2}\right)(E(S_2)-E(S_3)) + (b-d)E(S_3)\right\}. \tag{5.10}$$

The fraction N^{-1} is the frequency of A in the initial generation, while the expression in curly brackets represents its projected average excess in payoff. This is the sum of the differences between the marginal payoff to A and the mean payoff to a competitor in the same generation over all generations $t \geq 0$ as long as fixation is not reached.

The concept of projected average excess in payoff for A can be better understood with the help of Fig. 5. Consider a focal offspring (F) of type A in generation $t \geq 0$. We want to compare its marginal payoff to the mean payoff in the same generation. This mean will be the expected payoff to a typical offspring (T) chosen at random in the same generation. If this offspring has the same ancestor in generation 0 as the focal offspring, then its marginal payoff will also be the same. Therefore, it suffices to consider the case of distinct ancestors for F and T in generation 0. Then a third offspring (I) is chosen at random in the same generation and it may interact with either F or T.

Let S_3 be the number of generations backwards in time for the first coalescence event in the genealogies of the three offspring F, T, and I, and S_2 be the corresponding number for F and T only. If $t < S_3$, the three ancestors in generation 0 are all distinct and therefore T and I are both of type B. Then the payoff to F would be b compared to d for T. On the other hand, if $S_3 \leq t < S_2$ with F and I having a common ancestor in generation 0, whose conditional probability is $1/2$, then F and I are of type A, while T is of type B. This gives a payoff a to F compared to c to T. Finally, if $S_3 \leq t < S_2$ but with F and I having a common ancestor in generation 0, whose conditional probability is $1/2$, then T and I are of type B, while F is of type A. In this case, the payoff to F is b compared to d for T. In all other cases, F and T would be of the same type A, and then they would have the same payoff.

The final argument for the interpretation follows from the facts that

$$\sum_{t=0}^{\infty} P(S_3 > t) = E(S_3) \tag{5.11}$$

and

$$\sum_{t=0}^{\infty} P(S_3 \leq t < S_2) = \sum_{t=0}^{\infty} \left(P(S_2 > t) - P(S_3 > t)\right) = E(S_2) - E(S_3). \tag{5.12}$$

Scaled expected times in the limit of a large population size are obtained by multiplying S_2 and S_3 by c_N and by letting N tend to infinity, that is,

$$\mu_i = \lim_{N \to \infty} E(c_N S_i), \tag{5.13}$$

for $i = 2, 3$. Then the sign of the first derivative of the probability of ultimate fixation of A, and therefore whether or not weak selection favors A for replacement, is given by the sign of a scaled projected average excess in fitness.

Let us summarize.

Proposition 4. *In the case of a single initial A and in the limit of a large population size, the condition given in Propositions 2 and 3 for weak selection to favor A replacing B is equivalent to*

$$a_A = \left(\frac{a-c+b-d}{2}\right)(\mu_2 - \mu_3) + (b-d)\mu_3 > 0, \tag{5.14}$$

where μ_2 and μ_3 designate expected times, in number of c_N^{-1} generations in the limit of a large population size, with two and three lineages, respectively, and a_A represents a scaled projected average excess in payoff of A.

Note that

$$\mu_2 \geq 3\mu_3, \tag{5.15}$$

and the one-third law of evolution

$$x^* = \frac{d-b}{a-b-c+d} < \frac{1}{3} \tag{5.16}$$

is obtained when $\mu_2 = 3\mu_3$, which occurs with $\mu_2 = 1$ in the case of the Kingman coalescent.

6. Island model with dispersal preceding selection

In this section we examine the effect of a group structure on the condition for a single A to be favored for replacing B. Actually we consider the Wright (1931) island model for a population subdivided into a finite number of groups of the same size, assuming a Wright-Fisher reproduction scheme within groups and partial uniform dispersal of offspring before selection.

We have D groups of N parents producing virtually infinite numbers of offspring in equal relative proportions, that is, $(ND)^{-1}$ for each parent. We suppose that a fixed proportion m of offspring disperse uniformly among all groups, while the complementary proportion $1-m$ stay in their native group. This is followed by random pairwise interactions within groups affecting viability as previously. Finally N parents are sampled at random in each group to start the next generation.

Under the assumption of a Wright-Fisher reproduction scheme, the frequency of A in the offspring in group k in generation t before dispersal, for $k = 1,\ldots,D$ and $t \geq 0$, is the same as the frequency of A in the parents of group k at the beginning of generation t, denoted by $z_k(t)$. Then this frequency becomes

$$x_k(t) = (1-m)z_k(t) + m\overline{z(t)} \tag{6.1}$$

in the offspring after dispersal, and

$$\tilde{x}_k(t) = \frac{x_k(t)(1+sw_A(x_k(t)))}{1+sw(x_k(t))} \tag{6.2}$$

in the offspring after selection. Here,

$$\overline{z(t)} = D^{-1}\sum_{k=1}^{D} z_k(t) = D^{-1}\sum_{k=1}^{D} x_k(t) = \overline{x(t)} \tag{6.3}$$

is the frequency of A in all parents of generation t, which is the same as the frequency of A in all their offspring before dispersal as well as after dispersal, but before selection (see Fig. 6).

Proceeding as previously, we find that the probability of ultimate fixation of A is

$$\begin{aligned} u(s) &= E_s\left[\overline{z(\infty)}\right] \\ &= \overline{z(0)} + \sum_{t=0}^{\infty} E_s\left[\overline{z(t+1)} - \overline{z(t)}\right] \\ &= u(0) + D^{-1}\sum_{t=0}^{\infty}\sum_{k=1}^{D} E_s\left[\tilde{x}_k(t) - x_k(t)\right], \end{aligned} \tag{6.4}$$

where

$$E_s\left[\tilde{x}_k(t) - x_k(t)\right] = s(a-b-c+d)E\left[x_k(t)(1-x_k(t))(x_k(t)-x^*)\right] + o(s). \tag{6.5}$$

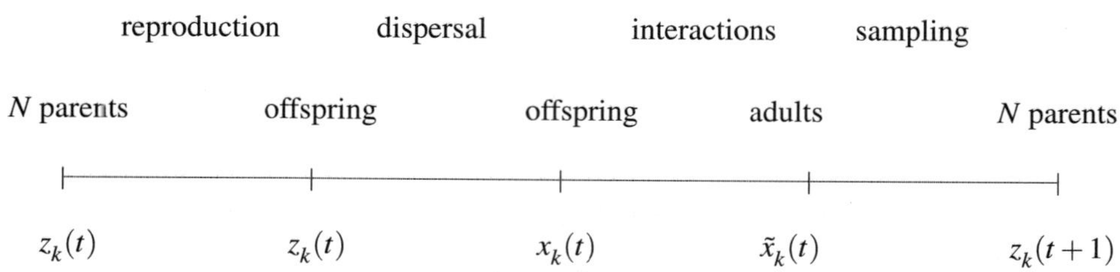

FIGURE 6. Life cycle from generation t to generation $t+1$ and notation for the frequency of A in group k at each step in the island model with dispersal before selection.

Actually the derivative evaluated at $s = 0$ is given by

$$u'(0) \quad = \quad (a-b-c+d)D^{-1}\sum_{t=0}^{\infty}\sum_{k=1}^{D} E\left[x_k(t)(1-x_k(t))(x_k(t)-x^*)\right]. \tag{6.6}$$

We conclude that $u'(0) > 0$ if and only if $x^* < \hat{x}$, where

$$\hat{x} = \frac{\sum_{t\geq 0} E[\overline{x(t)^2(1-x(t))}]}{\sum_{t\geq 0} E[\overline{x(t)(1-x(t))}]}. \tag{6.7}$$

Here, we have

$$\overline{x(t)^2(1-x(t))} = D^{-1}\sum_{k=1}^{D} x_k(t)^2(1-x_k(t)) \tag{6.8}$$

and

$$\overline{x(t)(1-x(t))} = D^{-1}\sum_{k=1}^{D} x_k(t)(1-x_k(t)). \tag{6.9}$$

Then the tower property of conditional expectation ascertains the following statement.

Proposition 5. *Consider the Wright island model for a finite number of groups of size N and assume a Wright-Fisher reproduction scheme followed by uniform dispersal of a proportion m of offspring and viability selection within groups according to the game of Proposition 1. Weak selection favors A replacing B if*

$$x^* < \frac{\sum_{t\geq 0} E[\xi_1(t)\xi_2(t)(1-\xi_3(t))]}{\sum_{t\geq 0} E[\xi_1(t)(1-\xi_2(t))]} = \hat{x}, \tag{6.10}$$

where $\xi_1(t), \xi_2(t), \xi_3(t)$ are indicator random variables for type A in offspring chosen at random without replacement in the same group chosen at random in generation t after dispersal.

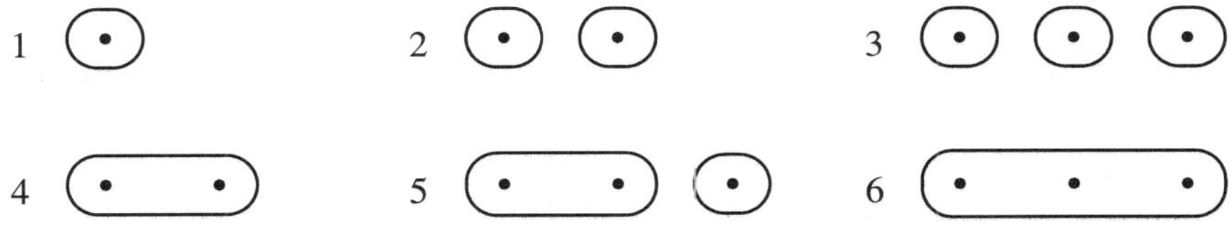

FIGURE 7. States for the ancestors of three offspring in the island model.

7. Calculation for the island model with dispersal preceding selection

We want to calculate $\hat{x}$ in Proposition 5 for the island model with dispersal preceding selection in the limit of a large number of groups and in the case where A is initially a single mutant. Without loss of generality, suppose $z_1(0) = N^{-1}$ and $z_k(0) = 0$ for $k = 2, \ldots, D$. See Ladret and Lessard (2007) for the analysis in the case of a fixed number of groups.

We will have to trace backwards in time the ancestors of two or three offspring after dispersal. Actually we will just need to know the number of groups d containing at least one ancestor and the number of groups n_i containing i ancestors for $i = 1, \ldots, d$ with $1 \leq \sum_{i=1}^{d} n_i \leq 3$. There are six possible states in the form $\mathbf{n} = (n_1, \ldots, n_d)$: (1), $(2,0)$, $(3,0,0)$, $(0,1)$, $(1,1,0)$, $(0,0,1)$, and they are labeled from 1 to 6 (see Fig. 7).

The state space S is partitioned into two subsets, $S_1 = \{1,2,3\}$ with all ancestors in different groups and $S_2 = \{4,5,6\}$ with at least two ancestors in the same group. State 1 is absorbing while all other states are transient. As D increases, transitions from the other states occur according to two different timescales with expected sojourn times in state 4, 5 or 6 becoming negligible compared to expected sojourn times in state 2 or 3.

As shown in Appendix A1, in the limit $D \to \infty$ with ND generations as unit of time, lineages within the same group either coalesce or migrate instantaneously to different groups, while each pair of lineages in different groups coalesces at rate f_{22}, which is the probability for two offspring chosen at random without replacement in the same group after dispersal to have ultimately two ancestors in different groups in the case of an infinite number of groups. In other words, after an initial scattering phase during which instantaneous transitions from states in S_2 to states in S_1 take place, there is a collecting phase during which transitions within S_1 occur according to the Kingman (1982) coalescent but with rate f_{22} instead of 1.

Let $p_{ij}(t)$ be the probability for the chain to be in state j and $v_{ij}(t)$ the probability for the chain to visit state j for the first time in the t-th generation backwards in time, given that the chain is in state i in the current generation. Note that

$$v_{ij} = \sum_{t \geq 1} v_{ij}(t) \tag{7.1}$$

is the probability for the chain to reach state j from state i for $j \neq i$. Moreover,

$$E(T_i) = (ND)^{-1} \sum_{t \geq 0} p_{ii}(t) \tag{7.2}$$

is the expected value of the time T_i spent in state i starting from state i before absorption into state 1 with ND generations as unit of time. In particular we have (see Appendix A1)

$$\lim_{D \to \infty} E(T_2) = f_{22}^{-1} \text{ and } \lim_{D \to \infty} E(T_4) = 0, \tag{7.3}$$

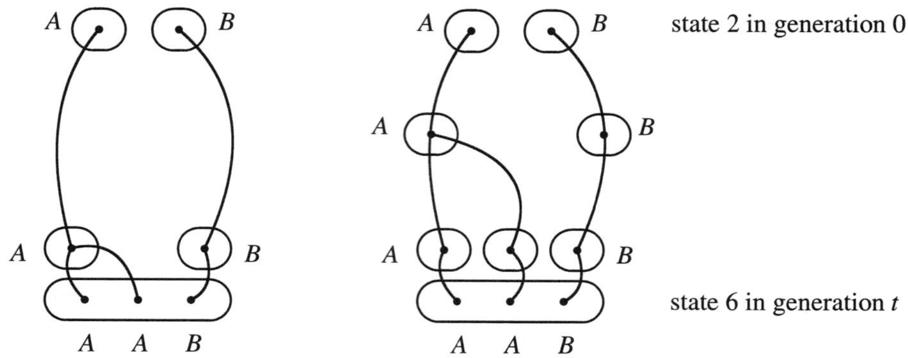

FIGURE 8. Lineages of three offspring of types A, A, B in the same group in the island model from generation t to generation 0.

so that only the time spent in state 2 has to be taken into account in the expected time with two lineages in the limit of a large population size. Moreover,

$$\lim_{D\to\infty} v_{42} = f_{22} = 1 - f_{21} \text{ and } \lim_{D\to\infty} v_{62} = f_{32} + f_{33} = 1 - f_{31}, \tag{7.4}$$

where f_{nk} represents the probability for n offspring chosen at random without replacement in the same group after dispersal to have ultimately k ancestors in different groups in the case of an infinite number of groups.

Considering all possible transitions from state 4 for two offspring chosen at random without replacement in generation $t \geq 0$ after dispersal to states in generation 0 so that the two offspring are of types A and B in this order, we obtain

$$\sum_{t\geq 0} E[\xi_1(t)(1-\xi_2(t))] = (ND)^{-1} \sum_{t\geq 1} p_{42}(t) + (ND)^{-1} \sum_{t\geq 1} p_{44}(t), \tag{7.5}$$

where

$$\sum_{t\geq 1} p_{42}(t) = \sum_{t\geq 1} \sum_{r=1}^{t} v_{42}(r) p_{22}(t-r) = \sum_{r\geq 1} \sum_{t\geq 0} v_{42}(r) p_{22}(t). \tag{7.6}$$

Owing to (7.1), (7.2), (7.3), (7.4), we conclude that

$$\sum_{t\geq 0} E[\xi_1(t)(1-\xi_2(t))] = v_{42} E(T_2) + E(T_4) - (ND)^{-1} \to 1,$$

as $D \to \infty$.

For three offspring chosen at random without replacement in state 6 in generation $t \geq 0$ after dispersal and of types A, A and B in this order, we obtain in a similar way

$$\sum_{t\geq 0} E[\xi_1(t)\xi_2(t)(1-\xi_3(t))] = (3ND)^{-1} \sum_{t\geq 1} p_{62}(t) + (3ND)^{-1} \sum_{t\geq 1} p_{64}(t), \tag{7.7}$$

from which

$$\sum_{t\geq 0} E[\xi_1(t)\xi_2(t)(1-\xi_3(t))] = \frac{v_{62}}{3} E(T_2) + \frac{v_{64}}{3} E(T_4) \to \frac{1-f_{31}}{3(1-f_{21})}, \tag{7.8}$$

as $D \to \infty$. Here, $1/3$ is the probability that two lineages in particular coalesce given that two lineages out of three coalesce (see Fig. 8).

Exact expressions of f_{21} and f_{31} in terms of m and N are given in Appendix A1. Note that the inequality $f_{31} < f_{21}$ always holds.

It remains to plug the above calculations in the upper bound given in Proposition 5. The following conclusion ensues.

Proposition 6. *In the case of a single initial A, the upper bound $\hat{x}$ in the condition given in Proposition 5 for weak selection to favor A replacing B in the island model with dispersal preceding selection satisfies*

$$\lim_{D\to\infty} \hat{x} = \frac{1-f_{31}}{3(1-f_{21})} > \frac{1}{3}, \tag{7.9}$$

where f_{21} and f_{31} are the probabilities that two and three offspring, respectively, chosen at random without replacement in the same group after dispersal have ultimately a common ancestor in the case of an infinite number of groups.

Proposition 6 means a less stringent condition for a single A to be favored for replacing B when the population is subdivided into a large number of small groups.

8. Island model with dispersal following selection

In this section we consider a variant of the previous island model by assuming that uniform dispersal occurs after selection. The main effect of this assumption is to introduce differential contributions of groups according to their composition from one generation to the next.

Here, the frequency of A in the offspring in group k in generation t goes from $x_k(t) = z_k(t)$ before selection to

$$\tilde{x}_k(t) = \frac{x_k(t)(1+sw_A(x_k(t)))}{1+sw(x_k(t))} \tag{8.1}$$

after selection, and finally to

$$\tilde{\tilde{x}}_k(t) = \frac{(1-m)x_k(t)(1+sw_A(x_k(t))+mD^{-1}\sum_{l=1}^{D} x_l(t)(1+sw_A(x_l(t))}{(1-m)(1+sw(x_k(t))+mD^{-1}\sum_{l=1}^{D}(1+sw(x_l(t))} \tag{8.2}$$

after selection and dispersal, since the relative size of group k after selection is $1+sw(x_k(t))$. (See Fig. 9.)

After some algebraic manipulations, the frequency of A in generation t in the whole population after selection and dispersal is found to be

$$\begin{aligned} D^{-1}\sum_{k=1}^{D}\tilde{\tilde{x}}_k(t) &= \overline{x(t)} + s(b-d)\overline{x(t)(1-x(t))} \\ &+ s(a-b-c+d)\overline{x(t)^2(1-x(t))} \\ &+ sm(2-m)(b+c-2d)(\overline{x(t)^2} - \overline{x(t)}^2) \\ &+ sm(2-m)(a-b-c+d)(\overline{x(t)^3} - \overline{x(t)}\ \overline{x(t)^2}) + o(s). \end{aligned} \tag{8.3}$$

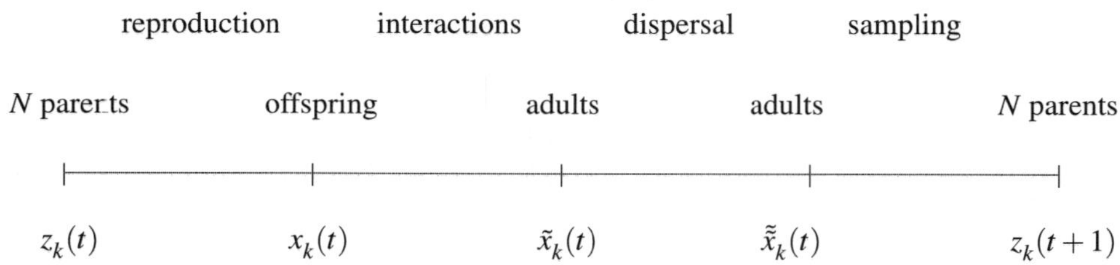

FIGURE 9. Life cycle from generation t to generation $t+1$ and notation for the frequency of A in group k at each step in the island model with dispersal after selection.

Here, $\overline{x(t)}$, $\overline{x(t)(1-x(t))}$ and $\overline{x(t)^2(1-x(t))}$ are defined as in Section 6, while

$$\begin{aligned}\overline{x(t)^2}-\overline{x(t)}^2 &= D^{-1}\sum_{k=1}^{D}x_k(t)^2-\left(D^{-1}\sum_{k=1}^{D}x_k(t)\right)^2\\ &= D^{-2}\sum_{k=1}^{D}\sum_{l=1,l\neq k}^{D}x_k(t)(1-x_l(t))-(1-D^{-1})\overline{x(t)(1-x(t))}\end{aligned}\tag{8.4}$$

and

$$\begin{aligned}\overline{x(t)^3}-\overline{x(t)}\ \overline{x(t)^2} &= D^{-1}\sum_{k=1}^{D}x_k(t)^3-\left(D^{-1}\sum_{k=1}^{D}x_k(t)\right)\left(D^{-1}\sum_{l=1}^{D}x_l(t)^2\right)\\ &= D^{-2}\sum_{k=1}^{D}\sum_{l=1,l\neq k}^{D}x_k(t)^2(1-x_l(t))-(1-D^{-1})\overline{x(t)^2(1-x(t))}.\end{aligned}\tag{8.5}$$

The tower property of conditional expectation yields

$$E[\overline{x(t)(1-x(t))}]=E[\zeta_1(t)(1-\zeta_2(t))]\tag{8.6}$$

and

$$E[\overline{x(t)^2(1-x(t))}]=E[\zeta_1(t)\zeta_2(t)(1-\zeta_3(t))]\tag{8.7}$$

as before, but with $\zeta_1(t),\zeta_2(t),\zeta_3(t)$ being indicator random variables for A in offspring chosen at random without replacement in generation t in the same group *before dispersal*.

Proceeding as in the previous section, we find that

$$\lim_{D\to\infty}\sum_{t\geq 0}E[\zeta_1(t)(1-\zeta_2(t))]=1\tag{8.8}$$

and

$$\lim_{D\to\infty}\sum_{t\geq 0}E[\zeta_1(t)\zeta_2(t)(1-\zeta_3(t))]=\frac{1-\tilde{f}_{31}}{3(1-\tilde{f}_{21})},\tag{8.9}$$

where

$$\tilde{f}_{n1}=f_{n1}(1-m)^{-n}\tag{8.10}$$

represents the probability that n offspring chosen at random without replacement in the same group *before dispersal* have ultimately a common ancestor in the case of an infinite number of groups.

On the other hand, we have

$$E\left[(D^2-D)^{-1}\sum_{k=1}^{D}\sum_{l=1,l\neq k}^{D} x_k(t)(1-x_l(t))\right] = E[\zeta_1(t)(1-\eta_2(t))] \tag{8.11}$$

and

$$E\left[(D^2-D)^{-1}\sum_{k=1}^{D}\sum_{l=1,l\neq k}^{D} x_k(t)^2(1-x_l(t))\right] = E[\zeta_1(t)\zeta_2(t)(1-\eta_3(t))], \tag{8.12}$$

where $\eta_2(t)$ and $\eta_3(t)$ are indicator random variables for A in offspring chosen at random without replacement in generation t *before dispersal*, but in a *different group* than the one for the indicator random variables $\zeta_1(t), \zeta_2(t), \zeta_3(t)$. In this case, we find that

$$\lim_{D\to\infty}\sum_{t\geq 0} E[\zeta_1(t)(1-\eta_2(t))] = \frac{1}{1-\tilde{f}_{21}} \tag{8.13}$$

and

$$\lim_{D\to\infty}\sum_{t\geq 0} E[\zeta_1(t)\zeta_2(t)(1-\eta_3(t))] = \frac{1}{3}+\frac{\tilde{f}_{21}}{1-\tilde{f}_{21}}. \tag{8.14}$$

These results are obtained by considering all transitions from states 2 and 5, respectively, for offspring sampled at random without replacement in generation $t \geq 0$ before dispersal to states in generation 0 that are compatible with the sample configuration.

The probability of ultimate fixation of A as a function of the intensity of selection is given by

$$u(s) = u(0) + D^{-1}\sum_{t=0}^{\infty}\sum_{k=1}^{D} E_s\left[\tilde{\tilde{x}}_k(t) - x_k(t)\right]. \tag{8.15}$$

Its derivative evaluated at $s=0$ is given by

$$\begin{aligned} u'(0) &= (b-d)\sum_{t\geq 0} E[\overline{x(t)(1-x(t))}] \\ &+ (a-b-c+d)\sum_{t\geq 0} E[\overline{x(t)^2(1-x(t))}] \\ &+ m(2-m)(b+c-2d)\sum_{t\geq 0} E[\overline{x(t)^2}-\overline{x(t)}^2] \\ &+ m(2-m)(a-b-c+d)\sum_{t\geq 0} E[\overline{x(t)^3}-\overline{x(t)}\ \overline{x(t)^2}]. \end{aligned} \tag{8.16}$$

In the limit of a large number of groups and after some algebraic manipulations, we find that

$$\begin{aligned} \lim_{D\to\infty} u'(0) &= (b-d)+(a-b-c+d)\frac{1-\tilde{f}_{21}+(1-m)^2(\tilde{f}_{21}-\tilde{f}_{31})}{3(1-\tilde{f}_{21})} \\ &+ (a-d)m(2-m)\frac{\tilde{f}_{21}}{1-\tilde{f}_{21}}. \end{aligned} \tag{8.17}$$

Using the exact expressions of $f_{21} = (1-m)^2 \tilde{f}_{21}$ and $f_{31} == (1-m)^3 \tilde{f}_{31}$ given in Appendix A1, it can be checked that

$$m(2-m)\frac{\tilde{f}_{21}}{1-\tilde{f}_{21}} = \frac{1}{N-1} \tag{8.18}$$

and

$$\frac{1-\tilde{f}_{21}+(1-m)^2(\tilde{f}_{21}-\tilde{f}_{31})}{3(1-\tilde{f}_{21})} > \frac{1-f_{31}}{3(1-f_{21})}, \tag{8.19}$$

as soon as $N > 1$. Then the condition $\lim_{D\to\infty} u'(0) > 0$ yields the following result.

Proposition 7. *In the case of dispersal following selection in the Wright island model of Proposition 5 in the limit of a large number of groups of fixed size $N > 1$, weak selection favors a single A replacing B if*

$$x^* < \frac{1-\tilde{f}_{21}+(1-m)^2(\tilde{f}_{21}-\tilde{f}_{31})}{3(1-\tilde{f}_{21})} + \frac{a-d}{(N-1)(a-b-c+d)}, \tag{8.20}$$

where $\tilde{f}_{21}$ and $\tilde{f}_{31}$ are the probabilities that two and three offspring, respectively, chosen at random without replacement in the same group before dispersal have ultimately a common ancestor in the case of an infinite number of groups.

Note that the upper bound for x^* given in Proposition 7 is always larger than the upper bound given in Proposition 6. This means an even less stringent condition for A to be favored for replacing B in the Wright island model when dispersal follows selection instead of preceding it.

9. Modified island model with skewed contributions of groups preceding selection

In this section we consider the effect of a skewed distribution for the contribution of a group in offspring in a subdivided population. We assume D groups of size N with dispersal of offspring preceding selection in each generation as in the island model of Section 6. However, with a small probability $D^{-\beta}$ for $\beta < 1$, one group chosen at random provides a proportion χ of all offspring, produced equally by all members of the group, compared to $(1-\chi)(D-1)^{-1}$ for every other group. With the complementary probability, the proportion is uniformly the same. In all cases, a proportion m of offspring in each group disperse and are replaced by as many migrants chosen at random among all migrants. This is followed by selection and random sampling within each group to start the next generation. This corresponds to the Eldon-Wakeley model applied to groups instead of parents.

The conclusion of Proposition 5 still holds. Moreover, a two-timescale argument can be applied in the limit of a large number of groups as in Section 7, but with ND^β generations as unit of time (see Appendix A2).

In number of ND^β generations, the expected time spent in state i before absorption into state 1 is written in the form

$$E(T_i) = (ND^\beta)^{-1} \sum_{t\geq 0} p_{ii}(t). \tag{9.1}$$

It can be shown that

$$\lim_{D\to\infty} E(T_2) = \lambda_{21}^{-1} \text{ and } \lim_{D\to\infty} E(T_4) = 0, \tag{9.2}$$

where λ_{21} represents the rate of coalescence of two lineages in different groups backwards in time in the limit of a large number of groups. Moreover, the limiting probabilities of reaching state 2 from states 4 and 6, respectively, are given by

$$\lim_{D\to\infty} v_{42} = f_{22} \text{ and } \lim_{D\to\infty} v_{62} = f_{32} + \frac{f_{33}\lambda_{32}}{\lambda_{32}+\lambda_{31}}, \tag{9.3}$$

where λ_{3i} represents the rate of transition from 3 to i lineages, for $i = 1,2$, in different groups backwards in time in the limit of a large number of groups.

Assuming a single initial A and using (9.1), (9.2), (9.3), we find that

$$\begin{aligned} D^{1-\beta}\sum_{t\geq 0} E[\xi_1(t)(1-\xi_2(t))] &= v_{42}E(T_2)+E(T_4)-(ND^{\beta})^{-1} \\ &\to f_{22}\lambda_{21}^{-1}, \end{aligned} \tag{9.4}$$

and

$$\begin{aligned} D^{1-\beta}\sum_{t\geq 0} E[\xi_1(t)\xi_2(t)(1-\xi_3(t))] &= \frac{v_{62}}{3}E(T_2)+\frac{v_{64}}{3}E(T_4) \\ &\to \frac{f_{32}}{3\lambda_{21}}+\frac{f_{33}\lambda_{32}}{3\lambda_{21}(\lambda_{32}+\lambda_{31})}, \end{aligned} \tag{9.5}$$

as $D \to \infty$. Here, $\xi_1(t), \xi_2(t), \xi_3(t)$ are indicator random variables for type A in offspring chosen at random without replacement in the same group chosen at random in generation t after dispersal like in Proposition 5.

This leads to the following result.

Proposition 8. *In the case of the Wright island model of Proposition 5 for D groups with a proportion m of migrant offspring each generation in each group before selection but a probability D^{β} for $\beta < 1$ that they come in proportion χ from a same group chosen at random, weak selection favors a single A replacing B in the limit of a large number of groups if*

$$x^* < \frac{1-f_{31}-f_{33}\left(\frac{\lambda_{31}}{\lambda_{32}+\lambda_{31}}\right)}{3(1-f_{21})} < \frac{1-f_{31}}{3(1-f_{21})}, \tag{9.6}$$

where f_{21} and f_{31} are defined as in Proposition 6, while λ_{31} and λ_{32} are the rates of transition from 3 to 1 and from 3 to 2, respectively, for the number lineages in different groups backwards in time with ND^{β} generations as unit of time in the limit of a large number of groups.

Proposition 8 means a more stringent condition for a single A to be favored for replacing B in an island model with a highly skewed distribution for the contribution of a group in the limit of a large number of groups.

10. Summary and comments

In conclusion we have shown in this paper that:

- Viability selection determined by the Iterated Prisoner's Dilemma (IPD) in an infinite population predicts the increase in frequency of Tit-for-Tat (A) against Always-Defect (B), and therefore can explain the spread of cooperation, but only from a frequency $x > x^{\star}$, where $x^{\star}$ is the frequency of A at an unstable polymorphic equilibrium.

- Weak viability selection determined by the IPD game in a finite population favors a single mutant A replacing B, and therefore can explain the advantage for cooperation to go to fixation from a low frequency, but only under the condition $x^\star < \hat{x}$ for some threshold frequency $\hat{x}$.

- In the limit of a large population size, we have $\hat{x} \leq 1/3$. Actually $\hat{x} = 1/3$, which is known as the one-third law of evolution, in a Wright-Fisher model, and more generally in the domain of application of the Kingman coalescent. On the other hand, $\hat{x} < 1/3$, which leads to a more stringent condition for the evolution of cooperation, if the contribution of a parent in offspring has a skewed enough distribution.

- In a group-structured population with uniform dispersal of offspring and weak viability selection within groups determined by the IPD game in the limit of a large number of groups of finite size, we have $\hat{x} > 1/3$. This means a less stringent condition for cooperation to evolve. Moreover, the condition is weaker if dispersal occurs after selection rather than before selection so that there are differential contributions of groups according to their composition. On the other hand, the condition is still weaker but to a lesser extent if the contribution of a group in offspring has a highly skewed distribution.

- The first-order effect of selection on the probability of fixation of a single mutant strategy is proportional to a projected average excess in payoff. This is the excess in payoff to the mutant strategy compared to the mean payoff in the population not only in the current generation but in all future generations as long as fixation is not reached.

Our results are based on approximations for the probability of ultimate fixation of a single mutant that are ascertained under the assumption of very weak selection. Actually, the intensity of selection is assumed to be small compared to the intensity of the other evolutionary forces. These are random drift, whose intensity is measured by the inverse of the population size, and dispersal in the case of a group-structured population, whose rate is supposed to be constant as the population size increases. On the other hand, the approach is not limited by restrictive assumptions on the production of offspring by parents or groups.

An alternative approach under the assumption that the intensity of selection is of the same order of magnitude as the other evolutionary forces is a diffusion approximation (see, e.g., Kimura 1984, Nagylaki 1980, 1997, Lessard 2005, 2007b, 2009). In this case, however, the contributions of parents and groups in offspring cannot be too highly skewed in distribution to avoid jump processes.

Our motivation in this paper was the evolution of cooperation and this is the reason for considering the Prisoner's Dilemma and its iterated version with Tit-for-Tat (A) and Always-Defect (B) as strategies. Of course, the approach used to deduce the first-order effect of selection on the probability of fixation of a single mutant is not limited to this particular game. Indeed, it does not depend on special relationships between the payoffs a, b, c and d. It holds for any two-player two-strategy matrix game with strategies A and B being the best replies to themselves or not.

Actually the approach is not limited to a matrix game, or linear expected payoffs $w_A(x)$ and $w_B(x)$ to A and B, respectively, with respect to the frequency of A represented by x. It can be extended to more general cases of frequency dependence with $w_A(x) - w_B(x)$ being a polynomial of any degree n with respect to x. Then expected backward times with up to

$n+2$ lineages have to be computed to approximate the fixation probability. Moreover, this can be used to get approximations in the case where the difference $w_A(x) - w_B(x)$ is any continuous function of x. (See Lessard and Ladret 2007.)

An approximation for the fixation probability can be obtained also in the case of a matrix game with any number of strategies. Then the approximation depends on the initial state of the population. Moreover, it can be expressed in terms of projected average excess in payoff given any initial frequencies (Lessard and Lahaie 2009).

We have considered pairwise interactions between offspring in infinite numbers. The case of pairwise interactions between adults in finite numbers is also of interest and it can be dealt with in a similar manner (see, e.g., Lessard 2005, Hilbe 2011). The analysis of the more general case of a multi-player game like the Public goods game is more recondite but not out of reach (Kurokawa and Ihara 2009, Gokhale and Traulsen 2010, Lessard 2011a).

Finally it can be shown that a matrix game in a finite group-structured population with uniform dispersal of offspring, or local extinction and recolonization, and payoff matrix A within groups is formally equivalent in the limit of a large population size to a matrix game in a well-mixed population with some effective game matrix A° (Lessard 2011b). The entries of this matrix are linear combinations of interaction or competition effects weighted by coefficients of identity-by-descent in an infinite population in the absence of selection. Then what is known about matrix games (see, e.g., Lessard 1990, Hofbauer and Sigmund 1998) can be applied *mutatis mutandis*.

Appendix A1. Two timescales for the Wright island model

Consider the neutral Wright island model for D groups of size N. In each generation, infinite numbers of offspring are produced in equal proportions and a fraction m of these disperse uniformly among all groups. This is followed by random sampling of N offspring in each group to start the next generation.

The six possible states for the ancestors of three offspring chosen after dispersal are given in Fig. 7. The transition matrix from one generation to the previous one, whose entries are represented by $p_{ij}(1)$ for i,j in $S = \{1,\ldots,6\}$, takes the form

$$\mathbf{P} = \mathbf{R} + (ND)^{-1}\mathbf{M}(D), \tag{10.1}$$

where $\mathbf{R}$ is the transition matrix in the case of an infinite number of groups. See Lessard and Wakeley (2004) for exact expressions of $\mathbf{R}$ and $\mathbf{M}(D)$.

Since all states in $S_1 = \{1,2,3\}$ are absorbing and all states in $S_2 = \{4,5,6\}$ transient in the case $D = \infty$, the ergodic theorem guarantees that

$$\lim_{t\to\infty} \mathbf{R}^t = \mathbf{H} = \begin{pmatrix} \mathbf{I} & \mathbf{0} \\ \mathbf{F} & \mathbf{0} \end{pmatrix}, \tag{10.2}$$

where $\mathbf{I}$ designates the 3×3 identity matrix and $\mathbf{0}$ the 3×3 zero matrix. Moreover,

$$\mathbf{F} = \begin{pmatrix} f_{21} & f_{22} & 0 \\ 0 & f_{21} & f_{22} \\ f_{31} & f_{32} & f_{33} \end{pmatrix}, \tag{10.3}$$

with f_{nk} denoting the probability for n offspring chosen at random without replacement in the same group after dispersal to have ultimately k ancestors in different groups in the case of an infinite number of groups. On the other hand, it can be checked that

$$\lim_{D\to\infty} \mathbf{M}(D) = \mathbf{M} = \begin{pmatrix} \mathbf{M}_{11} & \mathbf{M}_{12} \\ \mathbf{M}_{21} & \mathbf{M}_{22} \end{pmatrix}, \tag{10.4}$$

where

$$(10.5) \qquad \mathbf{M}_{11} = \begin{pmatrix} 0 & 0 & 0 \\ m(2-m) & -Nm(2-m) & 0 \\ 0 & 3m(2-m) & -3Nm(2-m) \end{pmatrix}$$

and

$$(10.6) \qquad \mathbf{M}_{12} = \begin{pmatrix} 0 & 0 & 0 \\ (N-1)m(2-m) & 0 & 0 \\ 0 & 3(N-1)m(2-m) & 0 \end{pmatrix}.$$

Applying a lemma due to Möhle (1998) to the transition matrix from time 0 to time τ in the past with ND generations as unit of time, we obtain

$$(10.7) \qquad \lim_{D\to\infty} \mathbf{P}^{\lfloor ND\tau \rfloor} = \mathbf{H}e^{\tau\mathbf{HMH}} = \begin{pmatrix} e^{\tau\mathbf{G}} & \mathbf{0} \\ \mathbf{F}e^{\tau\mathbf{G}} & \mathbf{0} \end{pmatrix} = \mathbf{Q}(\tau),$$

where $\lfloor\ \rfloor$ denotes the integer value and

$$(10.8) \qquad \mathbf{G} = \mathbf{M}_{11} + \mathbf{M}_{12}\mathbf{F} = f_{22} \begin{pmatrix} 0 & 0 & 0 \\ 1 & -1 & 0 \\ 0 & 3 & -3 \end{pmatrix}.$$

This uses the equality

$$(10.9) \qquad f_{22} = Nm(2-m)\left[\frac{1}{N} + \left(1 - \frac{1}{N}\right) f_{21}\right],$$

which can deduced from the exact expressions of f_{21} and $f_{22} = 1 - f_{21}$ (see below).

The matrix $\mathbf{G}$ is the generator of the death process of the Kingman (1982) coalescent with rate f_{22} instead of 1. The matrix $\mathbf{Q}(\tau)$, whose entries are denoted by $q_{ij}(\tau)$ for i, j in S, is a transition matrix from time 0 to time τ for a continuous-time Markov chain with initial instantaneous transitions from states in S_2 to states in S_1 and generator $\mathbf{G}$ for transitions within S_1.

The expected time in state 2 in number of ND generations is

$$(10.10) \qquad E(T_2) = (ND)^{-1} \sum_{t=0}^{\infty} p_{22}(t) = \int_0^\infty p_{22}(\lfloor ND\tau \rfloor) d\tau,$$

from which

$$(10.11) \qquad \lim_{D\to\infty} E(T_2) = \int_0^\infty q_{22}(\tau) d\tau = f_{22}^{-1}.$$

This is the case because two lineages coalesce at the rate f_{22} in the limit of a large number of groups. Moreover,

$$(10.12) \quad p_{22}(\lfloor ND\tau \rfloor) \leq \left(1 - \frac{m(2-m)}{ND}\right)^{\lfloor ND\tau \rfloor} \leq (1 - N^{-1})^{-1} e^{-m(2-m)\tau}.$$

Therefore, the dominated convergence theorem can be applied. Similarly the expected time in state 4 in number of ND generations is

$$(10.13) \qquad E(T_4) = (ND)^{-1} \sum_{t=0}^{\infty} p_{44}(t)$$

and

$$(10.14) \qquad \lim_{D\to\infty} E(T_4) = \int_0^\infty q_{44}(\tau) d\tau = 0,$$

since $q_{44}(\tau) = 0$ for all $\tau > 0$.

On the other hand, the vector $\mathbf{v}_{\bullet 2}^T = (0, 1, v_{32}, v_{42}, v_{52}, v_{62})$, where v_{i2} is the probability of reaching state 2 from state i for $i = 3, \ldots, 6$, satisfies the linear system of equations

$$\mathbf{v}_{\bullet 2} = \tilde{\mathbf{P}}^{ND} \mathbf{v}_{\bullet 2}, \tag{10.15}$$

where $\tilde{\mathbf{P}}$ is the transition matrix on S with state 2 assumed to be absorbing. In this case, Möhle's (1998) lemma yields

$$\lim_{D\to\infty} \tilde{\mathbf{P}}^{ND} = \tilde{\mathbf{Q}} = \begin{pmatrix} e^{\tilde{\mathbf{G}}} & \mathbf{0} \\ \mathbf{F}e^{\tilde{\mathbf{G}}} & \mathbf{0} \end{pmatrix}, \tag{10.16}$$

where

$$\tilde{\mathbf{G}} = f_{22} \begin{pmatrix} 0 & 0 & 0 \\ 0 & 0 & 0 \\ 0 & 3 & -3 \end{pmatrix}. \tag{10.17}$$

Therefore,

$$\lim_{D\to\infty} \mathbf{v}_{\bullet 2} = \tilde{\mathbf{Q}} \lim_{D\to\infty} \mathbf{v}_{\bullet 2}. \tag{10.18}$$

It can be checked directly that the unique solution is

$$\lim_{D\to\infty} \mathbf{v}_{\bullet 2}^T = (0, 1, 1, f_{22}, 1, f_{32} + f_{33}). \tag{10.19}$$

Finally, $f_{22} = 1 - f_{21}$ and $f_{32} + f_{33} = 1 - f_{31}$, where

$$f_{21} = (1-m)^2 \left[\frac{1}{N} + \left(1 - \frac{1}{N}\right) f_{21} \right], \tag{10.20}$$

$$f_{31} = (1-m)^3 \left[\frac{1}{N^2} + \frac{3}{N}\left(1 - \frac{1}{N}\right) f_{21} + \left(1 - \frac{1}{N}\right)\left(1 - \frac{2}{N}\right) f_{31} \right]. \tag{10.21}$$

This system of linear equations is obtained from a first-step analysis. Its solution is given by

$$f_{21} = \frac{(1-m)^2}{Nm(2-m) + (1-m)^2}, \tag{10.22}$$

$$f_{31} = f_{21} \left[\frac{N(1-m) + 2(N-1)(1-m)^3}{N^2 m(3 - 3m + m^2) + (3N-2)(1-m)^3} \right]. \tag{10.23}$$

Note that

$$f_{32} = 3(f_{21} - f_{31}). \tag{10.24}$$

This is the case because there are 3 possibilities for two offspring out of three to have a common ancestor.

Similarly the vector $\mathbf{v}_{\bullet 3} = (0, 0, 1, v_{43}, v_{53}, v_{63})$, where v_{i3} is the probability of reaching state 3 from state i for $i = 3, \ldots, 6$, must satisfy

$$\lim_{D\to\infty} \mathbf{v}_{\bullet 3} = \tilde{\tilde{\mathbf{Q}}} \lim_{D\to\infty} \mathbf{v}_{\bullet 3}, \tag{10.25}$$

where

$$\tilde{\tilde{\mathbf{Q}}} = \begin{pmatrix} \mathbf{I} & \mathbf{0} \\ \mathbf{F} & \mathbf{0} \end{pmatrix}. \tag{10.26}$$

The unique solution is

$$\lim_{D\to\infty} \mathbf{v}_{\bullet 3}^T = (0, 0, 1, 0, f_{22}, f_{33}). \tag{10.27}$$

Appendix A2. Two timescales for the modified Wright island model

Consider the neutral Wright island model for D groups of size N but suppose that, in each generation and with probability $D^{-\beta}$ for $\beta < 1$, the proportion of offspring produced equally by all members of a group chosen at random is χ compared to $(1-\chi)(D-1)^{-1}$ in every other group. With the complementary probability, the proportion is uniformly the same. In all cases, a proportion m of offspring in each group disperse and they are replaced by as many migrants chosen at random among all migrants before random sampling of N offspring to start the next generation.

The transition matrix on the state space for the ancestors of three offspring chosen after dispersal takes the form

$$\mathbf{P} = \mathbf{R} + (ND^{\beta})^{-1}\mathbf{M}(D), \tag{10.28}$$

where $\mathbf{R}$ is the same as in Appendix A1. The entries of the matrix $\mathbf{M}(D)$ can be found explicitly (Lasalle Ialongo 2008). The important point is that

$$\lim_{D\to\infty} \mathbf{M}(D) = \mathbf{M} = \begin{pmatrix} \mathbf{M}_{11} & \mathbf{M}_{12} \\ \mathbf{M}_{21} & \mathbf{M}_{22} \end{pmatrix}, \tag{10.29}$$

where

$$\mathbf{M}_{11} = \begin{pmatrix} 0 & 0 & 0 \\ (\chi m)^2 & -N(\chi m)^2 & 0 \\ N^{-1}(\chi m)^3 & 3(\chi m)^2(1-\chi m) & -3N(\chi m)^2 + 2N(\chi m)^3 \end{pmatrix}$$

and

$$\mathbf{M}_{12} = \begin{pmatrix} 0 & 0 & 0 \\ (N-1)(\chi m)^2 & 0 & 0 \\ 3(1-N^{-1})(\chi m)^3 & 3(N-1)(\chi m)^2(1-\chi m) & (1-N^{-1})(N-2)(\chi m)^3 \end{pmatrix}.$$

In this case, Möhle's (1998) lemma guarantees that

$$\lim D \to \infty \mathbf{P}^{\lfloor ND^{\beta}\tau \rfloor} = \begin{pmatrix} e^{\tau\mathbf{G}} & \mathbf{0} \\ \mathbf{F}e^{\tau\mathbf{G}} & \mathbf{0} \end{pmatrix} = \mathbf{Q}(\tau), \tag{10.30}$$

where

$$\mathbf{G} = \mathbf{M}_{11} + \mathbf{M}_{12}\mathbf{F} = \begin{pmatrix} 0 & 0 & 0 \\ \lambda_{21} & -\lambda_{21} & 0 \\ \lambda_{31} & \lambda_{32} & -\lambda_{31} - \lambda_{32} \end{pmatrix}. \tag{10.31}$$

The parameters λ_{lk} for $l > k \geq 1$ are the rates of transition from l to k lineages in different groups backwards in time with ND^{β} generations as unit of time as $D \to \infty$. We find that

$$\begin{aligned}
\lambda_{21} &= N(\chi m)^2 \left[\frac{1}{N} + \left(1-\frac{1}{N}\right) f_{21}\right], && (10.32)\\
\lambda_{31} &= N(\chi m)^3 \left[\frac{1}{N^2} + \frac{3}{N}\left(1-\frac{1}{N}\right) f_{21} + \left(1-\frac{1}{N}\right)\left(1-\frac{2}{N}\right) f_{31}\right], && (10.33)\\
\lambda_{32} &= N(\chi m)^3 \left[\frac{3}{N}\left(1-\frac{1}{N}\right) f_{22} + \left(1-\frac{1}{N}\right)\left(1-\frac{2}{N}\right) f_{32}\right], && (10.34)\\
&+ 3N(\chi m)^2(1-\chi m)\left[\frac{1}{N} + \left(1-\frac{1}{N}\right) f_{21}\right].
\end{aligned}$$

Note that

$$\lambda_{lk} = N \sum_{l \geq j \geq n \geq k-l+j \geq 1} \binom{l}{j} (\chi m)^j (1-\chi m)^{l-j} p_{jn} f_{n,k-l+j}, \tag{10.35}$$

where p_{jn} is the probability that j offspring chosen at random without replacement in the same group before dispersal have n parents, and f_{nk} is the probability that n parents chosen at random without replacement in the same group have ultimately k ancestors in different groups in the case $D = \infty$. The relationships between the parameters f_{nk} for $3 \geq n \geq k \geq 1$ exhibited in Appendix A1 lead to the expressions

$$\lambda_{21} = N \left(\frac{\chi m}{1-m} \right)^2 f_{21}, \tag{10.36}$$

$$\lambda_{31} = N \left(\frac{\chi m}{1-m} \right)^3 f_{31}, \tag{10.37}$$

$$\lambda_{32} = 3N \left(\frac{\chi m}{1-m} \right)^2 f_{21} - 3N \left(\frac{\chi m}{1-m} \right)^3 f_{31}. \tag{10.38}$$

Note that

$$\lambda_{lk} = N \sum_{l \geq j \geq k-l+j \geq 1} \binom{l}{j} (\chi m)^j (1-\chi m)^{l-j} \tilde{f}_{n,k-l-j}, \tag{10.39}$$

where $\tilde{f}_{nk}$ is the probability that n offspring chosen at random without replacement in the same group *before dispersal* have ultimately k ancestors in different groups in the case $D = \infty$.

Proceeding as previously, the expected time with two lineages in different groups in number of ND^β generations before coalescence satisfies

$$E(T_2) = (ND^\beta)^{-1} \sum_{t=0}^{\infty} p_{22}(t) \to \lambda_{21}^{-1} \tag{10.40}$$

as $D \to \infty$, while the corresponding expected time with two lineages in the same group, $E(T_4)$, tends to 0.

Finally the vector $\mathbf{v}_{\bullet 2}^T = (0, 1, v_{32}, v_{42}, v_{52}, v_{62})$, where v_{i2} is the probability of reaching state 2 from state i for $i = 3, \ldots, 6$, satisfies

$$\lim_{D \to \infty} \mathbf{v}_{\bullet 2} = \begin{pmatrix} e^{\tilde{\mathbf{G}}} & \mathbf{0} \\ \mathbf{F} e^{\tilde{\mathbf{G}}} & \mathbf{0} \end{pmatrix} \lim_{D \to \infty} \mathbf{v}_{\bullet 2}, \tag{10.41}$$

where

$$\tilde{\mathbf{G}} = \begin{pmatrix} 0 & 0 & 0 \\ 0 & 0 & 0 \\ \lambda_{31} & \lambda_{32} & -\lambda_{31} - \lambda_{32} \end{pmatrix}. \tag{10.42}$$

The solution is found to be

$$\lim_{D \to \infty} \mathbf{v}_{\bullet 2}^T = \left(0, 1, \frac{\lambda_{32}}{\lambda_{32} + \lambda_{31}}, f_{22}, f_{21} + \frac{f_{22}\lambda_{32}}{\lambda_{32} + \lambda_{31}}, f_{32} + \frac{f_{33}\lambda_{32}}{\lambda_{32} + \lambda_{31}} \right). \tag{10.43}$$

References

[1] AXELROD, R. (1984) *The Evolution of Cooperation*. New York: Basic Books.

[2] CANNINGS, C. (1974) The latent roots of certain Markov chains arising in genetics: a new approach. I. Haploid models. *Adv. Appl. Prob.* **6**, 260–290.

[3] ELDON, B. AND WAKELEY, J. (2006) Coalescent processes when the distribution of offspring number among individuals is highly skewed. *Genetics* **172**, 2621–2633.

[4] FISHER, R. A. (1930) *The Genetical Theory of Natural Selection*. Oxford: Clarendon.

[5] GOKHALE, C. S. AND TRAULSEN, A. (2010) Evolutionary games in the multiverse. *Proc. Natl. Acad. Sci. USA* **107**, 5500–5504.

[6] HILBE, C. (2011) Local replicator dynamics: A simple link between deterministic and stochastic models of evolutionary game theory. *Bull. Math. Biol.* DOI 10.1007/s11538-010-9608-2.

[7] HOFBAUER, J. AND SIGMUND, K. (1998) *Evolutionary Games and Population Dynamics*. Cambridge: Cambridge University Press.

[8] IMHOF, L. A. AND NOWAK, M. A. (2006) Evolutionary game dynamics in a Wright-Fisher process. *J. Math. Biol.* **52**, 667–681.

[9] KIMURA, M. (1984) Evolution of an altruistic trait through group selection as studied by the diffusion equation method. *IMA J. Math. Appl. Med. Biol.* **1**, 1–15.

[10] KINGMAN, J. F. C. (1982) The coalescent. *Stoch. Proc. Appl.* **13**, 235–248.

[11] KUROKAWA, S. AND IHARA, Y. (2009) Emergence of cooperation in public goods games. *Proc. Roy. Soc. B* **276**, 1379–1384.

[12] LADRET, V. AND LESSARD, S. (2007) Fixation probability for a beneficial allele and a mutant strategy in a linear game under weak selection in a finite island model. *Theor. Pop. Biol.* **72**, 409–425.

[13] LASALLE IALONGO, D. (2008) *Processus de coalescence dans une population subdivisée avec possibilité de coalescences multiples*. M.Sc. Thesis, Université de Montréal.

[14] LESSARD, S. (1990) Evolutionary stability: One concept, several meanings. *Theor. Pop. Biol.* **37**, 159–170.

[15] LESSARD, S. (2005) Long-term stability from fixation probabilities in finite populations: New perspectives for ESS theory. *Theor. Pop. Biol.* **68**, 19–27.

[16] LESSARD, S. (2007a) Cooperation is less likely to evolve in a finite population with a highly skewed distribution of family size. *Proc. Roy. Soc. B* **274**, 1861–1865.

[17] LESSARD, S. (2007b) An exact sampling formula for the Wright-Fisher model and a conjecture about the finite-island model. *Genetics* **177**, 1249-1254.

[18] LESSARD, S. (2009) Diffusion approximations for one-locus multi-allele kin selection, mutation and random drift in group-structured populations: a unifying approach to selection models in population genetics. *J. Math. Biol.* **59**, 659–696.

[19] LESSARD, S. (2011a) On the robustness of the extension of the one-third law of evolution to the multi-player game *Dyn Games Appl.* DOI 10.1007/s13235-011-0010-y.

[20] LESSARD, S. (2011b) Effective game matrix and inclusive payoff in group-structured populations. *Dyn Games Appl.* to appear.

[21] LESSARD, S. AND LADRET, V. (2007) The probability of fixation of a single mutant in an exchangeable selection model. *J. Math. Biol.* **54**, 721–744.

[22] LESSARD, S. AND LAHAIE, P. (2009) Fixation probability with multiple alleles and projected average allelic effect on selection. *Theor. Pop. Biol.* **75**, 266–277.

[23] LESSARD, S. AND WAKELEY, J. (2004) The two-locus ancestral graph in a subdivided population: convergence as the number of demes grows in the island model. *J. Math. Biol.* **48**, 275–292.

[24] MCNAMARA, J. M., BARTA, Z., HOUSTON, A. I. (2004) Variation in behaviour promotes cooperation in the Prisoner's Dilemma game. *Nature* **428**, 747–748.

[25] MÖHLE, M. (1998) A convergence theorem for Markov chains arising in population genetics and the coalescent with selfing. *Adv. Appl. Prob.* **30**, 493–512.

[26] MÖHLE, M. (2000) Total variation distances and rates of convergence for ancestral coalescent processes in exchangeable population models. *Adv. Appl. Prob.* **32**, 983–993.

[27] MÖHLE, M. AND SAGITOV, S. (2001) A classification of coalescent processes for haploid exchangable population models. *Ann. Appl. Probab.* **29**, 1547–1562.

[28] MORAN, P. A. P. (1958) Random processes in genetics. *Proc. Camb. Phil. Soc.* **54**, 60–71.

[29] NAGYLAKI, T. (1980) The strong-migration limit in geographically structured populations. *J. Math. Biol.* **9**, 101–114.

[30] NAGYLAKI, T. (1997) The diffusion model for migration and selection in a plant population. *J. Math. Biol.* **35**, 409–431.

[31] NOWAK, M. A., SASAKI, A., TAYLOR, C. AND FUDENBERG, D. (2004) Emergence of cooperation and evolutionary stability in finite populations. *Nature* **428**, 646–650.

[32] OHTSUKI, H., BORDALO, P. AND NOWAK, M. A. 2007 The one-third law of evolutionary dynamics. *J. Theor. Biol.* **249**, 289–295.

[33] PITMAN, J. (1999). Coalescents with multiple collisions. *Annals of Probability* **27,** 1870–1902.

[34] SAGITOV, S. (1999). The general coalescent with asynchronous mergers of ancestral lines. *Journal of Applied Probability* **36,** 1116–1125.

[35] ROUSSET, F. (2003) A minimal derivation of convergence stability measures. *J. Theor. Biol.* **221**, 665–668.

[36] WRIGHT, S. (1931) Evolution in Mendelian populations. *Genetics* **16**, 97–159.

DÉPARTEMENT DE MATHÉMATIQUES ET DE STATISTIQUE, UNIVERSITÉ DE MONTRÉAL, C.P. 6128 SUCCURSALE CENTRE-VILLE, MONTRÉAL, QUÉBEC H3C 3J7 CANADA

E-mail address: lessards@dms.umontreal.ca

Index

ISBN 978-0-8218-5326-9

PSAPM/69